Clinical Update on Inflammatory Disorders of the Gastrointestinal Tract

Frontiers of Gastrointestinal Research

Vol. 26

Series Editor

Markus M. Lerch Greifswald

Clinical Update on Inflammatory Disorders of the Gastrointestinal Tract

Volume Editors

Julia Mayerle Greifswald
Herbert Tilg Hall in Tirol/Innsbruck

25 figures, 3 in color and 18 tables, 2010

Basel · Freiburg · Paris · London · New York · Bangalore · Bangkok · Shanghai · Singapore · Tokyo · Sydney

Frontiers of Gastrointestinal Research
Founded 1975 by L. van der Reis, San Francisco, Calif.

Julia Mayerle
Klinik für Innere Medizin A
Klinikum der Ernst-Moritz-Arndt-
Universität Greifswald
Friedrich-Loeffler-Strasse 23A
DE-17475 Greifswald

Herbert Tilg
Bezirkskrankenhaus Hall in Tirol/
Innsbruck
Christian Doppler Research
Laboratory for Gut Inflammation
Medical University Innsbruck
Milser Strasse 10-12
AT-6060 Hall in Tirol/Innsbruck

Library of Congress Cataloging-in-Publication Data

Clinical update on inflammatory disorders of the gastrointestinal tract /
volume editors, Julia Mayerle, Herbert Tilg.
p. ; cm. -- (Frontiers of gastrointestinal research, ISSN 0302-0665
; v. 26)
Includes bibliographical references and indexes.
ISBN 978-3-8055-9294-9 (hardcover : alk. paper)
1. Digestive organs--Pathophysiology. 2. Inflammatory bowel diseases. 3.
Inflammation. I. Mayerle, Julia. II. Tilg, Herbert. III. Series:
Frontiers of gastrointestinal research, v. 26. 0302-0665 ;
[DNLM: 1. Gastrointestinal Diseases. 2. Inflammation. W1 FR946E v.26
2010 / WI 140 C6408 2010]
RC802.9.C65 2010
616.3'44--dc22

2009036720

Bibliographic Indices. This publication is listed in bibliographic services, including Current Contents®.

www.karger.com
Printed in Switzerland on acid-free and non-aging paper (ISO 9706) by Reinhardt Druck, Basel
ISSN 0302–0665
ISBN 978–3–8055–9294–9

Contents

Preface

Disorders of the digestive tract and the liver impose a significant economic and health burden on society. The US National Institutes of Health have recently completed a survey according to which digestive diseases account for 35 outpatient visits and 5 hospital days per 100 residents annually. The direct cost for medical expenses amount to USD 100 billion for digestive disorders and the indirect cost to an additional USD 44 billion [1]. While the magnitude of these expenses is on a par with a good-sized modern-day economic stimulus package, the disorders also have a high social cost. Ten percent of all deaths are attributed to digestive disorders. The numbers in Europe are thought to correspond to those in the USA, and by far the largest proportion of patients are thought to be affected by inflammatory disorders of the liver, the pancreas and the GI tract.

The good news is that research into inflammatory digestive disorders is showing results, with new insights from research constantly being brought to the bedside, and a reduction in disease burden and mortality has been achieved for a number of disorders.

Inflammatory diseases of the GI tract no longer include only infectious disorders (for which long-established anti-infective treatments are available and constantly being improved), but also a number of complex immunological disorders which are currently attracting much scientific attention. In this rapidly developing field, where biologically relevant signalling pathways were identified only in recent years, therapies that are directly based on these research findings are becoming available. A prominent example is the TNF-α blockade used in inflammatory bowel disease. In the field of gastrointestinal inflammation the term 'from bench to bedside' has become a reality.

This volume also covers emerging diseases such as microscopic colitis or non-alcoholic fatty liver disease that have only recently moved into the focus of scientific inquiry but which may have an unappreciated socio-economic impact.

Not all previously established treatment regimens have stood the test of time, and recent studies have questioned the evidence for using, for example, antibiotics, parenteral feeding or probiotics for patients with severe acute pancreatitis. Most pancreas experts were surprised to learn that the PROPATRIA trial on the use of probiotics in severe acute pancreatitis showed evidence for a harmful effect for a seemingly harmless therapy. The lesson from such negative studies is that controlled clinical trials should not only test novel treatment approaches but also challenge old assumptions about the standard of care.

The association between chronic inflammation and the development of cancer was recognized more than a century ago. As early as 1863 the German pathologist Rudolf Virchow reported the presence of leukocytes in neoplastic tissues and suggested a connection between inflammation and cancer. Nowadays clear associations have been shown between a variety of chronic inflammatory disorders such as Crohn's disease, ulcerative colitis, pancreatitis, hepatitis or *Helicobacter pylori*-associated gastritis and an increased cancer risk of affected patients.

This volume of the *Frontiers in Gastroenterology* series includes up-to-date reviews on the relevant issues in inflammatory disorders of the GI tract, the liver and the pancreas. In a combination of expert basic research reviews and cutting-edge treatment guidelines the reader will learn about newly identified treatment targets and be able to participate in the development of novel treatment strategies. The fact that cancer often emerges on a background of inflammation highlights the notion that treating or preventing inflammation can also result in a reduction of cancer prevalence and is often effective in not only alleviating the patient's suffering but also in reducing mortality.

We are grateful that world-leading experts in several fields have agreed to contribute to this project and want to thank them for sharing their knowledge and expertise with our readers. We hope that you will find it as fascinating and instructive to read this book as we found working on it. We also hope that this volume may serve as an inspiration for clinicians and scientist to enter the rapidly developing field of inflammatory diseases in gastroenterology.

Julia Mayerle and *Herbert Tilg*
August 2009

Reference

1 Everhart J (ed): The Burden of Digestive Diseases in the United States. US Department of Health and Human Services, Public Health Service, National Institutes of Health, National Institute of Diabetes and Digestive and Kidney Diseases. Washington, US Government Printing Office, 2008. Available online at: www2.niddk.nih.gov/AboutNIDDK/ReportsAndStrategicPlanning/BurdenOfDisease/DigestiveDiseases (accessed July 27, 2009).

Mayerle J, Tilg H (eds): Clinical Update on Inflammatory Disorders of the Gastrointestinal Tract.
Front Gastrointest Res. Basel, Karger, 2010, vol 26, pp 1–14

Non-Alcoholic Fatty Liver Disease

Elisabetta Bugianesi

Department of Internal Medicine, Division of Gastro-Hepatology, San Giovanni Battista Hospital, University of Turin, Turin, Italy

Abstract

Non-alcoholic fatty liver disease (NAFLD) embraces a wide range of metabolic hepatic injuries that are characterized by steatosis, and it currently represents the most common liver disease in Western countries. NAFLD prevalence in the general population ranges between 3 and 30%, and between 10 and 15% of patients with NAFLD meet the current diagnostic criteria for non-alcoholic steatohepatitis (NASH). The long-term hepatic prognosis of NAFLD patients depends on their histological stage at diagnosis. Simple steatosis has a favourable outcome, whereas patients presenting with NASH can develop cirrhosis and hepatocellular carcinoma. NAFLD is commonly associated with the features of the metabolic syndrome and results from a complex interaction between multiple genes and environmental causes, with insulin resistance as the underlying mechanism. The factors responsible for progression from simple steatosis to steatohepatitis are still elusive, but lipotoxicity, oxidative stress and adipokine imbalance play pivotal roles. NAFLD is often asymptomatic and most patients present with incidentally found abnormal liver blood tests and/or 'bright liver' at ultrasound. Due to the absence of a distinctive serological marker, the process of diagnosing NAFLD is one of excluding other causes. Liver biopsy is the only reliable tool to identify NASH, but non-invasive markers of liver damage are being developed. In addition to treating relevant co-existing conditions, such as obesity, dyslipidemia and diabetes, a number of other strategies are being evaluated. These include insulin sensitizers, antioxidants, anti-cytokines and cytoprotective agents, angiotensin-receptor antagonists and glutathione precursors, but their efficacy remains low, and whether this is accompanied by an improvement in liver histology remains to be determined.

Nomenclature

The term 'non-alcoholic fatty liver disease' (NAFLD) encompasses a wide spectrum of metabolic liver injuries that are associated with over-accumulation of fat in the liver. It is morphologically indistinguishable from alcoholic fatty liver disease (AFLD), but occurs in subjects who do not consume a significant amount of alcohol. Liver histology ranges from simple steatosis (>5% fat infiltration, with/without minimal inflammation) to non-alcoholic steatohepatitis (NASH), which is characterized

Table 1. Conditions associated with NAFLD

Conditions
Metabolic syndrome (central obesity, impaired fasting glucose/type 2 diabetes, dyslipidemia, hypertension)
Polycystic ovary syndrome
Obstructive sleep apnoea
Familial and acquired lipodystrophies
Drugs (tamoxifen, amiodarone, highly active anti-retroviral therapy)
Jejunoileal bypass
Jejunal diverticulosis (contaminated bowel syndrome)
Massive intestinal resection
Malnutrition, cachexia
Total parenteral nutrition
Hypobetalipoproteinemia

by hepatocyte injury (ballooning degeneration and/or Mallory bodies), inflammation and/or fibrosis [1]. Pathological classification is not completely defined yet, but recently a new scoring system has been proposed [2]. Four histological features (steatosis, lobular inflammation, hepatocellular ballooning and fibrosis) were considered relevant to construct a NAFLD activity score used to classify cases into 'NASH', 'borderline' and 'not NASH'. Simple steatosis is thought to be a relatively benign state, whereas NASH represents the form of NAFLD that has the potential to progress to cirrhosis and hepatocellular carcinoma. The threshold for alcohol consumption that can reliably distinguish NAFLD from AFLD is still controversial, but most current studies use the cutoff of 70 g/week for women and 140 g/week for men [1]. NAFLD clusters with obesity, diabetes and is now commonly considered the hepatic manifestation of the metabolic syndrome (MS) [3]. Other conditions associated with NAFLD are referred to as 'secondary NAFLD' and are usually semantically linked to their aetiology (table 1).

Epidemiology and Natural History

In Western countries NAFLD currently represents the most common liver disease and is steadily increasing along with the worldwide spreading of obesity and diabetes; nevertheless, accurate estimates of prevalence, incidence and natural history are lacking. Available epidemiological data are biased by lack of sensitivity and specificity of the test used for the diagnosis (abnormal liver enzymes and/or hepatic ultrasound). Estimating the prevalence of NASH is even more problematic since the diagnosis requires liver biopsy. Based on liver enzymes, the likely prevalence of

NAFLD in the United States population is between 3 and 23% [4], similar to that reported in surveys using hepatic ultrasound (US) [5]. However, a recent study using proton magnetic resonance spectrometry (^{H}MRS) found that approximately 30% of the population have increased liver fat [6], although aminotransferases were normal in 80% of cases. About 10–15% of the patients with NAFLD meet the current diagnostic criteria for NASH, making the prevalence of NASH in the general population between 2 and 3% [7]. NAFLD occurrence increases with age, is generally higher in men than in pre-menopausal women and varies with ethnicity. The whole spectrum of NAFLD mostly occurs in patients with obesity (60–95%), type II diabetes mellitus (28–55%) and hyperlipidemia (27–92%) [3]. NAFLD was found in 86% of patients undergoing bariatric surgery, with fibrosis in 74%, and mild necro-inflammation in 24% of cases, while a post-mortem study reported NASH in 3% of lean, 19% of obese and 50% of a morbidly obese individuals [7]. The pattern of fat distribution is more important than BMI, and visceral fat has been associated with severity of inflammation and fibrosis [8]. Ultrasonographic evidence of 'bright' liver is nearly the rule in patients with type 2 diabetes, with a prevalence of 70% reported from an US survey [9]. Although no systematic study of liver biopsy has ever been performed, liver disease may be an important cause of death in diabetes [10]. The criteria for the MS are fulfilled in 18% of normal weight and 67% of obese non-diabetic NAFLD patients [11].

Studies in children have reported a prevalence of NAFLD of 3% in the general paediatric population and 53% in obese children [12]. Of relevance is the association between small gestational age at birth and NAFLD during childhood and adolescence [13].

The natural history of NAFLD is difficult to assess because most studies are retrospective, while prospective ones have not been running long enough to evaluate late complications. The overall survival of patients with NAFLD is less than that of a matched population, liver disease being the third leading cause of death in NAFLD patients compared to the 13th in a general population [14]. The long-term hepatic prognosis of NAFLD patients depends on the histological stage at diagnosis [15]. Over 8–13 years, 12–40% of patients with simple steatosis will develop NASH, while 15% of patients presenting with NASH will develop cirrhosis, increasing to 25% of patients with precirrhotic stage at diagnosis. Weight gain and advanced fibrosis are the most important risk factors for NAFLD progression [15]. Of note, steatosis progressively disappears as fibrosis develops and such cases present as cryptogenic cirrhosis. Up to 70% of patients with cryptogenic cirrhosis show clinical features suggestive of NASH. About 7% of subjects with NASH-related cirrhosis will develop a hepatocellular carcinoma within 10 years, while 50% will require transplantation or will die from liver-related causes [7].

The long-term natural history of subjects with NAFLD is affected by the presence of the underlying MS and the risk for liver disease is outweighed by the risk of diabetes and cardiovascular disease.

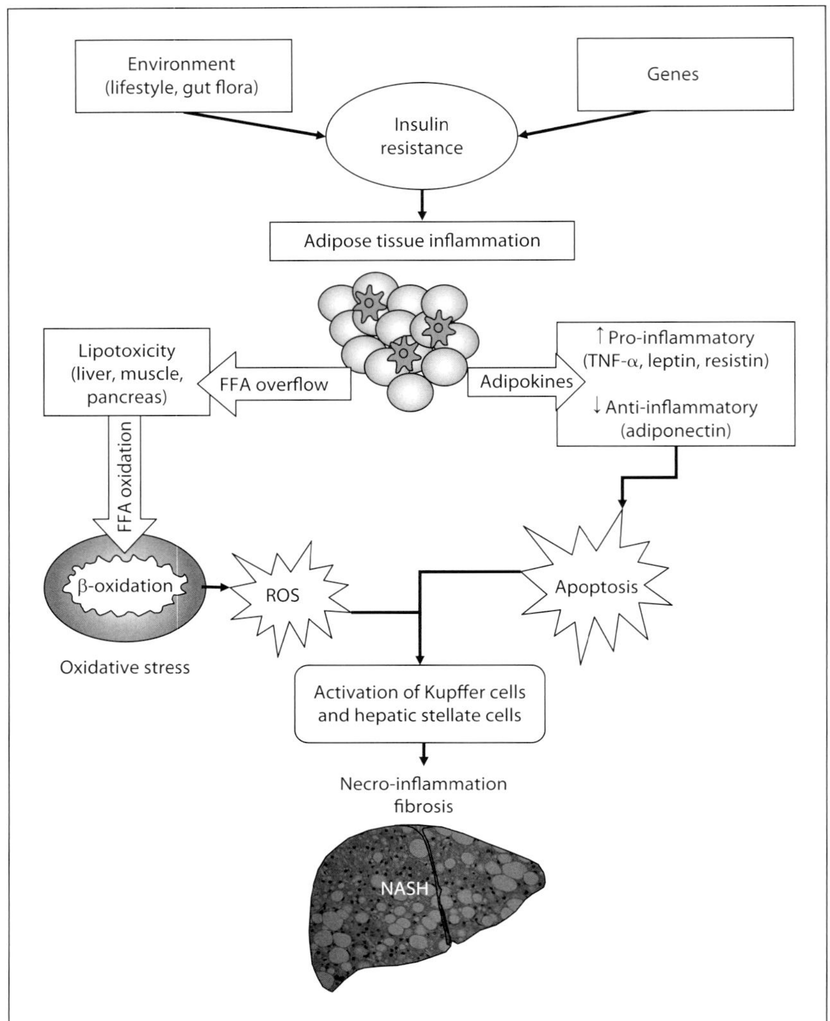

Fig. 1. Pathophysiological mechanisms of nonalcoholic fatty liver disease and nonalcoholic steatohepatitis. FFA = Free fatty acids; NASH = non-alcoholic steatohepatitis; ROS = reactive oxygen species.

Pathogenesis

NAFLD is a complex trait resulting from the interaction between multiple genes and social, behavioural and environmental factors (fig. 1). Among acquired factors, overeating and obesity (particularly visceral obesity), play a crucial role in the development of NAFLD. NAFLD patients have a higher intake of saturated fat, foods with high glycaemic index and soft drinks. The daily intake of refined sugars has been correlated with the extent of inflammatory changes at biopsy [7]. Upon specific dietary conditions, changes in gut microbiota can affect fat storage and energy harvesting and can also trigger an inflammatory response by increasing intestinal permeability and endotoxin absorption [16].

The different prevalence of NAFLD among racial groups suggests that genes play a role in the pathogenesis and natural history. In the Dallas Heart Study, the prevalence of NAFLD in Hispanic and African-Americans was 3-fold higher and 4-fold lower, respectively, compared with European-American patients [6]. Family studies reported the co-existence of NASH and/or cryptogenic cirrhosis in siblings and found that 18% of patients with NASH had an affected first degree relative [7].

The pathophysiological hallmark of NAFLD is insulin resistance in target tissues (liver, muscle and adipose tissue). Elevated free fatty acid (FFA) levels during fasting are constant findings in NAFLD patients and stem from accelerated lipolysis, the immediate result of insulin resistance in adipose tissue [3]. The influx of plasma FFA from fat stores, particularly from visceral fat, represents the major supply of intrahepatic triglycerides (62–82%) [17]. Other important sources are represented by hepatic de novo lipogenesis and by dietary intake, which respectively account for 25 and 15% of liver TG [17].

Factors responsible for the progression from simple fatty liver to NASH still remain elusive.

Lipotoxicity appears to be a key factor in the progression to steatohepatitis and is attributed to products of excessive oxidative metabolism of FFA by mithocondria, peroxisomes and microsomal enzymes, that induce elevated production of reactive oxygen species and other toxic intermediates, cell injury and programmed cell death [18]. Reactive oxygen species-mediated lipid peroxidation generates 4-hydroxynonenal and malondialdehyde that can stimulate the synthesis of extracellular matrix by hepatic stellate cells, ultimately leading to fibrosis. The role of steatosis per se, once considered the main culprit of the progression to NASH, needs to be reassessed in view of the recent findings that triglycerides represent a non-toxic form of lipid accumulation and might represent a protective mechanism from the cytotoxicity of FFA [19].

Advanced fibrotic liver disease is constantly associated with multiple features of the MS [11]. Insulin-resistant states are characterized by chronic subclinical inflammation, induced by an imbalance between pro-inflammatory (TNF-α, IL-6, leptin) and anti-inflammatory (adiponectin) adipokines released by an inflamed adipose tissue [3]. Circulating levels and hepatic expression of TNF-α are increased in NAFLD. TNF-α can interfere with insulin signalling and activate Kupffer cells, contributing to fibrosis. By contrast, adiponectin plasma levels are decreased and inversely related to hepatic insulin resistance, hepatic fat content, degree of inflammation and extent of fibrosis [3]. High TNF-α and low adiponectin plasma levels have been indicated as independent predictors of NASH in NAFLD patients [20].

All the mechanisms discussed above are capable of inducing apoptosis, currently considered the major mode of cell death in NASH. Induction of the pro-apoptotic pathway is mediated by up-regulation of Fas, activation of Jun N-terminal kinase and successive destabilization of lysosomes with release of cathepsin B, activation of NF-kB, increased transcription of TNF-α and, finally, mitochondrial dysfunction [21].

Clinical Features and Investigation

NAFLD is rarely perceived by the patient as an health problem, but NASH may have an asymptomatic course to overt liver disease. Therefore, an early diagnosis of NAFLD and the recognition of patients at risk for NASH are particularly important. The absence of a distinctive serological marker for the identification of NAFLD and the presence of normal liver enzyme in the majority (80%) of subjects render this task particularly challenging [6]. Importantly, there is no difference in histological severity between patients with and without abnormal tests [22]. The most common modes of presentation of NAFLD are detection of unexplained abnormal liver enzymes and/or of bright liver at US. Most patients are asymptomatic or complain about non-specific symptoms, such as fatigue, sleep disturbances or right upper quadrant discomfort. After exclusion of other causes of chronic liver disease, including excessive alcohol intake, NAFLD should be suspected in any individuals with 1 or more components of the MS. Diagnostic workup should include anthropometric measurements (BMI, waist circumference) and assessment of blood pressure. Hepatomegaly is the most common physical finding. More advanced liver disease is associated with signs of portal hypertension. Features of polycystic ovary syndrome (hyperandrogenism) should be sought in young women with suspected NAFLD. Liver function tests display mild (2- to 5-fold) elevations of transaminases, alkaline phosphatase and gamma glutamyltranspeptidase, but may be normal in the majority of NAFLD subjects. The alanine transaminase/aspartate transaminase ratio is <1 unless advanced fibrotic NAFLD is present or the patient has covert AFLD. Gamma glutamyltranspeptidase level is not discriminatory between ALFD and NAFLD, as raised levels are commonly associated with metabolic disease. Laboratory tests should cover the complete liver biochemistry, including platelets, albumin and coagulation. Lipid profile should be assessed, including apolipoprotein B levels, since hypobetalipoproteinemia is a rare, familial cause of NAFLD. Ferritin may be increased in up to 60% of patients, but is mostly an expression of subclinical inflammation, since iron overload is uncommon (found in 4–6% of NAFLD) [7]. Autoantibodies (anti-nuclear antibody and smooth muscle antibody) are often present at low titres and may be related with more advanced disease [7]. A test of insulin sensitivity is mandatory in all patients. The Homeostatic Model Assessment (HOMA-R) is easily obtained from fasting plasma glucose and insulin levels, but a 2-hour oral glucose load provides more information about the glucose tolerance status [3].

Liver imaging can give supporting evidence of steatosis by US, MR imaging or CT scan, but the sensitivity and specificity of these tests is poor where liver fat is

<33%. ^{H}MRS reliably detects even minimal amounts of steatosis (2–3%), although it is expensive and is mainly used for research purposes [6].

While the diagnosis of NAFLD can be derived from classical risk factors, along with US detection of hepatic steatosis, the most relevant challenge to the clinician is the distinction between simple fatty liver and NASH. In the absence of overt cirrhosis, no imaging modality can identify necro-inflammatory changes and fibrosis. Liver biopsy is the only reliable tool for the staging of the disease, but its widespread use is contraindicated due to a doubtful risk-benefit ratio. Consequently, different non-invasive approaches have been attempted. Predictive indices of disease severity are based on components of the MS along with biochemical/imaging indicators of advanced liver disease. Recently, a NAFLD fibrosis score has been developed that includes age, BMI, aspartate transaminase/alanine transaminase ratio, albumin, platelets and impaired fasting glucose/diabetes [23]. This score, combined with the European Liver Fibrosis panel of serum fibrosis markers, has an accuracy of >90% in identifying different stages in NAFLD. Plasma levels of caspase-generated cytokeratin-18 fragments, a marker of hepatocyte apoptosis, have also been used to identify NASH [24].

Another non-invasive procedure to detect severe fibrosis and cirrhosis is transient elastography (Fibroscan®; Echosens, Paris, France), but steatosis and obesity can limit its reliability. Although potentially interesting, all non-invasive surrogates of histological severity require further validation before they can be used in routine clinical practice.

With these considerations in mind, a reasonable approach to NAFLD patients is to consider liver biopsy according to figure 2.

Therapy

Debate continues on the most appropriate treatment for NAFLD because few large randomized controlled trials (RCTs) with histological end-points have been published (table 2). Therapeutic approaches are mainly focused on modification of risk factors [25]. Since a patient's lifestyle can likely affect the efficacy of any pharmacological compounds, it appears reasonable to start the management of NAFLD with formal diet and exercise. Currently, a low-calorie, low-fat diet is recommended for weight reduction in clinical practice. There is general agreement that lifestyle changes reduce aminotransferase levels, but very few data are available on histology. The initial target of weight loss should be 5–10% of baseline weight. More rapid weight loss (>1.5 kg/week) might promote histological exacerbation of NASH due to massive fatty acid mobilization from visceral stores. The most important limitation to lifestyle changes remains the patient's compliance; specific programs of cognitive behavioural therapy should be considered in non-compliant subjects. Anti-obesity treatment might be of help in selected patients when lifestyle modification is unsuccessful, but its efficacy

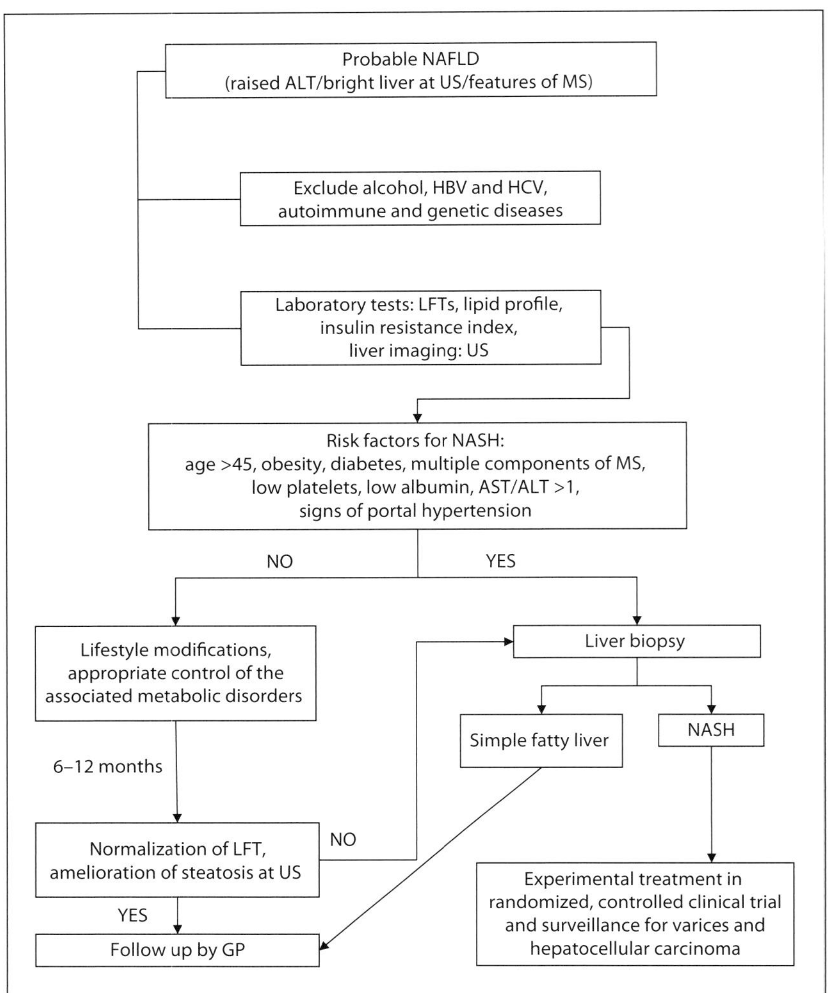

Fig. 2. Diagnostic flow chart of NAFLD/NASH. After having excluded other causes of chronic liver disease, patients with suspected NAFLD should undergo evaluation for components of the metabolic syndrome. Liver biopsy should be restricted to patients with at least some of the risk factors for advanced fibrosis or, after an initial attempt to normalize liver enzymes by lifestyle intervention, in whom this is not achieved after 6–12 months. Patients with simple steatosis can be managed by general practitioners, whereas patients with more advanced NAFLD require long-term follow-up by hepatologists in light of the need for surveillance for complications including esophageal varices and hepatocellular carcinoma. These patients will also be candidates for emerging therapies in large randomized controlled trials. ALT = Alanine transaminase; AST = aspartate transaminase; GPs = General practitioners; HBV = hepatitis B virus; HCC = hepatocellular carcinoma; HCV = hepatitis C virus; LFTs = liver function tests; MS = metabolic syndrome; US = ultrasound.

Table 2. Therapeutic trials of NAFLD with histological end-points

Reference	Experimental therapy	Control treatment	Patients n	Study duration	Effects on biopsy	
					steatosis	fibrosis
Lifestyle changes						
Andersen et al. [26]	diet		41	4–23 months	improved	variable
Ueno et al. [27]	diet, exercise	no treatment	25	3 months	improved	improved
Huang et al. [28]	diet		23	12 months	improved	no change
Insulin-sensitizers						
Nair et al. [29]	metformin		15	12 months	variable	variable
Uygun et al. [30]	metformin	diet	36	6 months	improved	no change
Bugianesi et al. [31]	metformin	diet, vitamin E	55	12 months	improved	improved
Caldwell et al. [32]	troglitazone		10	≤6 months	variable	variable
Promrat et al. [33]	pioglitazone		18	48 weeks	improved	improved
Sanyal et al. [34]	pioglitazone + vitamin E	vitamin E	20	6 months	improved	improved
Neuschwander-Tetri et al. [35]	rosiglitazone		30	48 weeks	improved	improved
Belfort et al. [36]	pioglitazone + diet	placebo + diet	55	6 months	improved	no change
Ratziu et al. [37]	rosiglitazone	placebo			improved	no change
Bariatric surgery						
Kral et al. [38]	biliopancreatic diversion		104	6–111 months	improved	worsened
Luyckx et al. [39]	gastroplasty		69	27 months	improved	no change
Dixon et al. [40]	gastric banding		36	26 months	improved	improved
Barker et al. [41]	gastric bypass		19	13-32 months	improved	improved
Mattar et al. [42]	gastric bypass		70	15 ± 9 months	improved	improved
Clark et al. [43]	gastric bypass		16	10 ± 4 months	improved	improved
Klein et al. [44]	gastric bypass		7	12 months	improved	no change
Lipid-lowering agents						
Laurin et al. [45]	clofibrate	UDCA	16	12 months	no change	no change
Anti-hypertensive agents						
Yokohama et al. [46]	losartan		7	48 weeks	no change	improved

Table 2. Continued

Reference	Experimental therapy	Control treatment	Patients n	Study duration	Effects on biopsy	
					steatosis	fibrosis
Cytoprotective agents						
Laurin et al. [45]	UDCA	clofibrate	24	12 months	improved	no change
Lindor et al. [47]	UDCA	placebo	166	24 months	no change	no change
Anti-oxidants						
Hasegawa et al. [48]	vitamin E	diet	22	12 months	variable	variable
Yoneda et al. [49]	vitamin E	diet		12 months	no change	improved
Harrison et al. [50]	vitamins E and C + diet	placebo + diet	45	6 months	no change	improved
Abdelmalek et al. [51]	betaine		8	12 months	improved	improved
Various treatments						
Harrison et al. [52]	orlistat		3	6 months	variable	variable

UDCA = Ursodeoxycholic acid.

in NAFLD has to be proven. Endocannabinoids antagonists, which produce a dose-dependent reduction in food intake interacting with anorexic and orexigenic pathways within the central nervous system, appeared promising, but rimonabant use has been associated with an increased incidence of severe mood-related disorders. Bariatric surgery is reserved to morbidly obese patients or in presence of major comorbidities. Biliopancreatic diversion should be avoided, whereas gastric banding and gastric bypass have shown encouraging results, even though a number of questions, such as durability and postoperative care, remain to be answered [25].

None of the drugs that can potentially be used in addition to diet and exercise (table 2) have been formally approved worldwide for treatment of NAFLD and/or NASH.

Drugs Targeting Components of the Metabolic Syndrome

Given the pivotal role of insulin resistance in NAFLD, treatment with insulin-sensitizing agents has a sound rationale. Metformin improves hepatic insulin resistance by down-regulating hepatic glucose production and diverting fatty acids to mitochondrial β-oxidation. In NAFLD trials, metformin has shown mixed results. In the largest RCT published so far, metformin treatment was associated with higher rates of aminotransferase normalization and with a significant decrease in liver fat, necro-inflammation and fibrosis [26]. Lactic acidosis was not observed in patients with severe fibrosis, but it may be a concern in decompensated cirrhosis. Treatment withdrawal is accompanied by a return of aminotransferase to pre-treatment levels. The National Institutes of Health is currently undertaking 2 large clinical trials (PIVENS and TONIC) to validate these preliminary results.

The novel class of peroxisome proliferator activated receptor γ agonists, thiazolidinediones, has also been tested in NAFLD. They shift fat accumulation from ectopic sites (muscle, liver) to adipose tissue by increasing plasma adiponectin levels, and they have anti-inflammatory effects. Recently, the first placebo-controlled trial of pioglitazone in patients with NASH showed a reversal of the metabolic milieu favouring steatosis and an amelioration of all the histologic features of steatohepatitis with the exception of fibrosis [27]. Pioglitazone treatment for NASH has to be continued long-term since its suspension led to a worsening of steatosis and inflammation. Rosiglitazone showed mixed results in early trials but failed to ameliorate necro-inflammation and fibrosis in the most recent RCT [28]. Weight gain, decreased haemoglobin and fluid retention are significant side effects of therapy with thiazolidinediones. Of note, several meta-analyses of trials in type 2 diabetes patients have shown that rosiglitazone increases the incidence of myocardial infarction and heart failure [7]. Since any therapy shown to be effective in NAFLD should be maintained long term or lifelong, a careful assessment of the risk-benefit ratio and drug safety profile is of paramount importance.

Dyslipidemia is an important component of the MS and is related to NAFLD. Although lipid-lowering agents (fibrates, statins) were effective in reducing aminotransferase levels in NAFLD, the evidence of a benefit in histological features is scanty [25]. However, their use appears to be safe and should be part of the treatment of MS according to guidelines.

Drugs Targeting Pathways of Liver Damage

A number of strategies targeting hypothetical mechanisms of hepatocellular damage are being evaluated, but the rationale for these therapeutic options is less sound. They include antioxidants (vitamin E), anti-cytokines (pentoxifylline) and cytoprotective (ursodeoxycholic acid) agents, angiotensin-receptor antagonists (losartan) and glutathione precursors (betaine) [25].

Vitamin E has shown beneficial effects in controlled trials performed in the paediatric NAFLD population, but in adults there is no evidence that antioxidant therapy is better than lifestyle changes. Betaine, pentoxifylline and losartan improved liver function tests in open label studies.

After promising preliminary data, the beneficial effect of ursodeoxycholic acid on liver histology was not confirmed in a large RCT [29]. Phlebotomy has been reported to improve hepatic histology in NAFLD patients and to ameliorate insulin resistance in patients with impaired glucose tolerance, despite normal body iron stores [30].

Although these drugs can normalize liver enzymes, whether this is accompanied by an improvement in liver histology remains to be determined and their routinely use is not recommend, yet.

Orthotopic Liver Transplant
Patients with NASH-related cirrhosis or those who develop hepatocellular carcinoma are candidates for liver transplantation. After orthotopic liver transplant, hepatic steatosis develops universally in cryptogenic as well as NASH-related cirrhosis, and after 2–5 years NASH may recur. However, graft function is maintained over the first 5–10 years after transplant and the rate of graft loss is not increased [7].

Conclusions

Only a few years ago, NAFLD was not considered a harmful disease and no specific treatment was indicated. A better knowledge of its natural history raised much interest on pathogenesis and treatment. Important tasks in the years to come are to assess the burden of liver morbidity and mortality in high-risk subgroups, to weigh the relative importance of cardiovascular versus liver complications in the final prognosis, and to develop non-invasive diagnostic tools and an effective, well-tolerated treatment.

References

1 Neuschwander-Tetri BA, Caldwell SH: Nonalcoholic steatohepatitis: summary of an AASLD single topic conference. Hepatology 2003;37:1202–1219.

2 Kleiner DE, Brunt EM, Van Natta M, et al: Design and validation of a histological scoring system for non-alcoholic fatty liver disease. Hepatology 2005;41:1313–1321.

3 Bugianesi E, McCullough AJ, Marchesini G: Insulin resistance: a metabolic pathway to chronic liver disease. Hepatology 2005;42:987–1000.

4 Ruhl CE, Everhart JE: Elevated serum alanine aminotransferase and gamma-glutamyltransferase and mortality in the United States population. Gastroenterology 2009;136:477–485.

5 Bedogni G, Miglioli L, Masutti F, Tiribelli C: Prevalence and risk factors for non-alcoholic fatty liver disease: the Dionysos nutrition and liver study. Hepatology 2005;42:44–52.

6 Browning JD, Szczepaniak LS, Dobbins R, et al: Prevalence of hepatic steatosis in an urban population in the United States: impact of ethnicity. Hepatology 2004;40:1387–1395.

7 De Alwis NMW, Day CP: Non-alcoholic fatty liver disease: the mist gradually clears. J Hepatol 2008;48:S104–S112.

8 Van Der Poorten D, Milner KL, Hui J, Hodge A, Trenell MI, Kench LG, London R, Peduto T, Chisholm DJ, George J. Visceral fat: a key mediator of steatohepatitis in metabolic liver disease. Hepatology 2008;48:449–457.

9 Targher G, Bertolini L, Padovani R, Rodella S, Tessari R, Zenari L, et al: Prevalence of non-alcoholic fatty liver disease and its association with cardiovascular disease among type 2 diabetic patients. Diabetes Care 2007;30:1212–1218.

10 de Marco R, Locatelli F, Zoppini G, Verlato G, Bonora E, Muggeo M: Cause-specific mortality in type 2 diabetes: the Verona Diabetes Study. Diabetes Care 1999;22:756–761.

11 Marchesini G, Bugianesi E, Forlani G, Cerrelli F, Lenzi M, Manini R, Natale S, Vanni E, Villanova N, Melchionda N, Rizzetto M: Nonalcoholic fatty liver, steatohepatitis, and the metabolic syndrome. Hepatology 2003;37:917–923.

12 Franzese A, Vajro P, Argenziano A, Puzziello A, Iannucci M, Saviano M, et al: Liver involvement in obese children: ultrasonography and liver enzyme levels at diagnosis and during follow up in an Italian population. Dig Dis Sci 1997;42:1428–1432.

13 Nobili V, Marcellini M, Marchesini G, Vanni E, Manco M, Villani A, Bugianesi E: Intrauterine growth retardation, insulin resistance, and nonalcoholic fatty liver disease in children. Diabetes Care 2007;30:2638–2640.
14 Adams LA, Lymp JF, St Sauver J, Sanderson SO, Lindor KD, Feldstein A, et al: The natural history of non-alcoholic fatty liver disease: a population-based cohort study. Gastroenterology 2005;129:113–121.
15 Ekstedt M, Franzen LE, Mathiesen UL, Thorelius L, Holmqvist M, Bodemar G, Kechagias S: Long-term follow-up of patients with NAFLD and elevated liver enzymes. Hepatology 2006;44:865–873.
16 Backhed F, Ding H, Wang T, Hooper LV, Koh GY, Nagy A, et al: The gut microbiota as an environmental factor that regulates fat storage. Proc Natl Acad Sci USA 2004;101:15718–15723.
17 Donnelly KM, Smith CI, Schwarzenberg SJ, Jessurum J, Boldt MD, Parks EJ: Sources of fatty acids stored in liver and secreted via lipoproteins in patients with nonalcoholic fatty liver disease. J Clin Invest 2005;115:1343–1351.
18 Malhi H, Gores GJ: Molecular mechanisms of lipotoxicity in nonalcoholic fatty liver disease. Semin Liver Dis 2008;28:360–369.
19 Yamaguchi K, Yang L, McCall S, Huang J, Yu XX, Pandey SK, et al: Inhibiting triglyceride synthesis improves hepatic steatosis but exacerbates liver damage and fibrosis in obese mice with nonalcoholic steatohepatitis. Hepatology 2007;45:1366–1374.
20 Hui JM, Hodge A, Farrell GC, Kench JG, Kriketos A, George J: Beyond insulin resistance in NASH: TNF-alpha or adiponectin? Hepatology 2004;40:46–54.
21 Li Z, Berk M, McIntyre TM, Gores GJ, Feldstein AE: The lysosomal-mitochondrial axis in free fatty acid-induced hepatic lipotoxicity. Hepatology 2008;47:1495–1503.
22 Mofrad P, Contos MJ, Haque M, et al: Clinical and histologic spectrum of nonalcoholic fatty liver disease associated with normal ALT values. Hepatology 2003;37:1286–1292.
23 Angulo P, Hui JM, Marchesini G, Bugianesi E, George J, Farrell GC, et al: The NAFLD fibrosis score: a non-invasive system that accurately identifies liver fibrosis in patients with NAFLD. Hepatology 2007;45:846–854.
24 Wieckowska A, Zein NN, Yerian LM, Lopez AR, McCullough AJ, Feldstein AE: In vivo assessment of liver cell apoptosis as a novel biomarker of disease severity in nonalcoholic fatty liver disease. Hepatology 2006;44:27–33.
25 Bugianesi E, Marzocchi R, Villanova N, Marchesini G: Non-alcoholic fatty liver disease/non-alcoholic steatohepatitis (NAFLD/NASH): treatment. Best Pract Res Clin Gastroenterol 2004;18:1105–1116.
26 Bugianesi E, Gentilcore E, Manini R, Natale S, Vanni E, Villanova N, David E, Rizzetto M, Marchesini G: A randomized controlled trial of metformin versus vitamin E or prescriptive diet in nonalcoholic fatty liver disease. Am J Gastroenterol 2005;100:1082–1090.
27 Ueno T, Sugawara H, Sujaku K, et al: Therapeutic effects of restricted diet and exercise in obese patients with fatty liver. J Hepatol 1997;27:103–107.
28 Huang MA, Greenson JK, Chao Z, et al: One year intense nutritional counseling results in histological improvement in patients with nonalcoholic steatohepatitis: a pilot study. Am J Gastroenterol 2005;100:1072–1081.
29 Nair S, Diehl AM, Wiseman M, et al: Metformin in the treatment of non-alcoholic steatohepatitis: a pilot open label trial. Aliment Pharmacol Ther 2004;20:23–28.
30 Uygun A, Kadayifci A, Isik AT, et al: Metformin in the treatment of patients with non-alcoholic steatohepatitis. Aliment Pharmacol Ther 2004;19:537–544.
31 Bugianesi E, Gentilcore E, Manini R, Natale S, Vanni E, Villanova N, David E, Rizzetto M, Marchesini G: A randomized controlled trial of metformin versus vitamin E or prescriptive diet in nonalcoholic fatty liver disease. Am J Gastroenterol 2005;100:1082–1090.
32 Caldwell SH, Hespenheide EE, Redick JA, et al: A pilot study of a thiazolidinedione, troglitazone, in nonalcoholic steatohepatitis. Am J Gastroenterol 2001;96:519–525.
33 Promrat K, Lutchman G, Uwaifo GI, et al: A pilot study of pioglitazone treatment for nonalcoholic steatohepatitis. Hepatology 2004;39:188–196.
34 Sanyal AJ, Mofrad PS, Contos MJ, et al: A pilot study of vitamin E versus vitamin E and pioglitazone for the treatment of nonalcoholic steatohepatitis. Clin Gastroenterol Hepatol 2004;2:1107–1115.
35 Neuschwander-Tetri BA, Brunt EM, Wehmeier KR, et al: Improved nonalcoholic steatohepatitis after 48 weeks of treatment with the PPAR-gamma ligand rosiglitazone. Hepatology 2003;38:1008–1017.
36 Belfort R, Harrison SA, Brown K, et al: A placebo-controlled trial of pioglitazone in subjects with nonalcoholic steatohepatitis. N Engl J Med 2006;355:2297–2307.

37 Ratziu V, Giral P, Jacqueminet S, Charlotte F, Hartemann-Heurtier A, Serfaty L, Podevin P, Lacorte JM, Bernhardt C, Bruckert E, Grimaldi A, Poynard T: LIDO study group. Rosiglitazone for nonalcoholic steatohepatitis: one-year results of the randomized placebo-controlled Fatty Liver Improvement with Rosiglitazone Therapy (FLIRT) Trial. Gastroenterology 2008;135:100–110.

38 Kral JG, Thung SN, Biron S, Hould FS, Lebel S, Marceau S, Simard S, Marceau P: Effects of surgical treatment of the metabolic syndrome on liver fibrosis and cirrhosis. Surgery 2004;135:48–58.

39 Luyckx FH, Scheen AJ, Desaive C, Thiry A, Lefebvre PJ: Parallel reversibility of biological markers of the metabolic syndrome and liver steatosis after gastroplasty-induced weight loss in severe obesity. J Clin Endocrinol Metab 1999;84:4293.

40 Dixon JB, Bhathal PS, Hughes NR, O'Brien PE: Nonalcoholic fatty liver: improvement in liver histological analysis with weight loss. Hepatology 2004; 39:1647–1654.

41 Barker KB, Paleka NA, Bowers SP, et al: Nonalcoholic steatohepatitis: effect of Roux-en-Y gastric bypass surgery. Am J Gastroenterol 2006;101:368–373.

42 Mattar SG, Velcu LM, Robinovitz M, et al: Surgically induced weight loss significantly improves nonalcoholic fatty liver disease and the metabolic syndrome. Ann Surg 2005;242:610–620.

43 Clark JM, Alkhuraishi AR, Solga SF, Alli P, Diehl AM, Magnuson TH: Roux-en-Y gastric bypass improves liver histology in patients with non-alcoholic fatty liver disease. Obes Res 2005;13:1180–1186.

44 Klein S, Mittendorfer B, Eagon C, et al: Gastric bypass surgery improves metabolic and hepatic abnormalities associated with nonalcoholic fatty liver disease. Gastroenterology 2006;130:1564–1572.

45 Laurin J, Lindor KD, Crippin JS, et al: Ursodeoxycholic acid or clofibrate in the treatment of non-alcohol-induced steatohepatitis: a pilot study. Hepatology 1996;23:1464–1467.

46 Yokohama S, Yoneda M, Hane DA, et al: Therapeutic efficacy of angiotensinogen II antagonist in patients with nonalcoholic steatohepatitis. Hepatology 2004; 40:1222–1225.

47 Lindor KD, Kowdley KV, Heathcote EJ, et al: Ursodeoxycholic acid for treatment of nonalcoholic steatohepatitis: results of a randomized trial. Hepatology 2004;39:770–778.

48 Hasegawa T, Yoneda M, Nakamura K, et al: Plasma transforming growth factor-beta1 level and efficacy of alpha-tocopherol in patients with non-alcoholic steatohepatitis: a pilot study. Aliment Pharmacol Ther 2001;15:1667–1672.

49 Yoneda M, Hasegawa T, Nakamura K, et al: Vitamin E therapy in patients with NASH. Hepatology 2004; 39:568.

50 Harrison SA, Torgerson S, Hayashi P, Ward J, Schenker S: Vitamin E and vitamin C treatment improves fibrosis in patients with nonalcoholic steatohepatitis. Am J Gastroenterol 2003;98:2485–2490.

51 Abdelmalek MF, Angulo P, Jorgensen RA, et al: Betaine, a promising new agent for patients with nonalcoholic steatohepatitis: results of a pilot study. Am J Gastroenterol 2001;96:2711–2717.

52 Harrison SA, Ramrakhiani S, Brunt EM, et al: Orlistat in the treatment of NASH: a case series. Am J Gastroenterol 2003;98:926–930.

53 Valenti L, Fracanzani AL, Dongiovanni P, Bugianesi E, Marchesini G, Manzini P, Vanni E, Fargion S: Iron depletion by phlebotomy improves insulin resistance in patients with nonalcoholic fatty liver disease and hyperferritinemia: evidence from a case-control study. Am J Gastroenterol 2007;102: 1251–1258.

Elisabetta Bugianesi, MD, PhD
UOADU Gastro-Epatologia, Università di Torino
Azienda Ospedaliera San Giovanni Battista
Corso Bramante 88, IT–10126 Torino (Italy)
Tel. +39 011 633 6397, Fax +39 011 633 5927, E-Mail ebugianesi@yahoo.it

Mayerle J, Tilg H (eds): Clinical Update on Inflammatory Disorders of the Gastrointestinal Tract.
Front Gastrointest Res. Basel, Karger, 2010, vol 26, pp 15–31

Fibrosis in the GI Tract: Pathophysiology, Diagnosis and Treatment Options

Massimo Pinzani

Dipartimento di Medicina Interna, Center for Research, Higher Education and Transfer DENOthe, Università degli Studi di Firenze, Florence, Italy

Abstract

A fibrogenic process mainly consequent to reiterated tissue damage is characteristic of different chronic disease affecting the liver, the pancreas and the intestine. Although the general mechanisms leading to fibrosis in different tissues of the gastro-intestinal tract are similar, the development and the consequences and fibrosis are specific for each affected organ. A chronic wound healing reaction is in general the main fibrogenic mechanism and is characterized by the simultaneous presence of inflammation, tissue remodelling and regeneration. In this context, activated myofibroblasts represent the main effectors of tissue fibrosis. In addition to local tissue myofibroblasts, other local mesenchymal cells such as fibroblasts, vascular pericytes (i.e. stellate cells) and smooth muscle cells can differentiate in activated myofibroblasts upon chronic damage. Myofibroblasts may also derive from epithelial or endothelial cells in processes termed epithelial-mesenchymal transition and endothelial-mesenchymal transition, respectively. In addition, a population of unique circulating fibroblast-like cells derived from bone marrow stem cells, commonly termed 'fibrocytes', has been shown to potentially contribute to the fibrogenic process. Several mechanisms involved in the fibrogenic process are outlined in this article, including the regulation of myofibroblast recruitment, proliferation, survival and pro-fibrogenic activity, the pro-fibrogenic role of innate and adaptative immune mechanisms, the role of oxidative stress, and the close association that occurs between fibrogenesis and angiogenesis. In addition, information on the clinical evaluation of fibrosis progression/regression and potential anti-fibrogenic approaches is also provided.

Cellular and Molecular Mechanisms of Fibrosis in the GI Tract

Several chronic diseases of the gastrointestinal tract are characterized by progressive tissue fibrosis leading to severe clinical complications that include organ failure and death.

In all these diseases, the fibrogenic process is mainly subsequent to the activation of a chronic wound healing reaction in response to a persistent irritant causing reiterated tissue damage. Although the general mechanisms leading to fibrosis in different tissues of the GI are the same, the development and the consequences and fibrosis are specific to the liver, the pancreas and the intestine.

Wound Healing versus Fibrosis

The chronic wound healing reaction is characterized by the simultaneous presence of inflammation, tissue remodelling and regeneration [1]. The deposition of fibrillar extracellular matrix (ECM) represents the best available solution aimed at maintaining tissue continuity in a context of extensive tissue necrosis. In general, newly deposited fibrillar ECM is rapidly degraded and tissue fibrosis is usually observed after a significant amount of time, when the rate of synthesis of fibrillar collagens (I, III, VI, etc.) by myofibroblasts exceeds the rate of degradation. This occurs for 2 main reasons: (1) the number of activated myofibroblasts reaches a peak hyperplasia partly because of a progressive resistance to apoptosis; and (2) the perpetuation of the activation of this cell type is characterized by a progressive reduction of its ability to degrade and remodel fibrillar ECM. In clinical terms, although moderate tissue fibrosis is usually not associated with significant clinical signs or decreased organ function, the presence of fibrosis is itself an important indicator in prognostic terms since it highlights the transition from effective wound repair to the fibrogenic evolution of the disease and represents a hallmark of chronically evolving disease. Figure 1 illustrates the possible outcomes of wound healing: tissue regeneration or fibrotic healing.

Effectors of Fibrogenesis: Myofibroblasts

Activated myofibroblasts are the main effectors of tissue fibrosis [2]. The term 'activated' is important to define the biologic features of these mesenchymal cells in disease conditions. Normally, myofibroblasts are key components of tissue stroma and play a key role in ECM homeostasis. In conditions of acute or chronic tissue damage, myofibroblasts undergo a process of activation that leads to their proliferation and migration to the area of damage where they reconstitute the ECM milieu necessary for tissue regeneration. This process is characterized by sequential steps: deposition of fibrillar ECM, scar contraction, degradation of fibrillar ECM and reconstitution of the normal tissue ECM. In case of chronic damage, there is an overlapping of the different phases of the wound healing process with a progressive accumulation of fibrillar ECM and, in this context, a key element is the perpetuation of myofibroblast activation [3]. In addition to enhanced cell proliferation and migration, chronically activated myofibroblasts are characterized by increased contractility, decreased sensibility to pro-apoptotic stimuli, secretion of pro-fibrogenic, pro-inflammatory and pro-angiogenic cytokines.

A key current issue concerns the cellular origin of myofibroblasts. It is well established that, in addition, to local tissue myofibroblasts, other mesenchymal cells present in the tissue can differentiate in activated myofibroblasts upon chronic damage. These include fibroblasts, vascular pericytes and smooth muscle cells (fig. 1). In

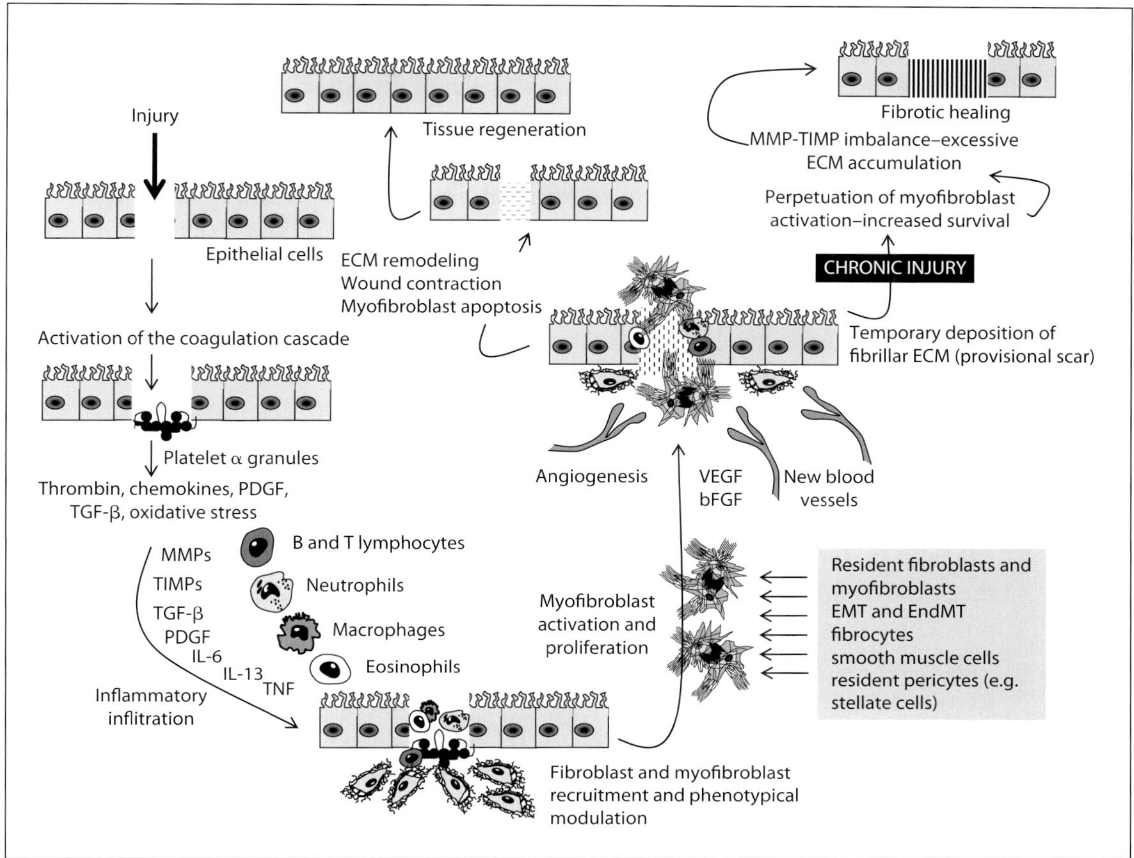

Fig. 1. The chronic wound healing/fibrogenic process. bFGF = Basic fibroblast growth factor; ENT = epithelial-mesenchymal transition; EndMT = endothelial-mesenchymal transition; PDGF = platelet-derived growth factor.

addition to resident mesenchymal cells, myofibroblasts may derive from epithelial or endothelial cells in processes termed epithelial-mesenchymal transition [4, 5] and endothelial-mesenchymal transition [6], respectively. More recently, a population of unique circulating fibroblast-like cells derived from bone marrow stem cells, commonly termed 'fibrocytes', has been identified and characterized. Fibrocytes express CD34, CD45 and type I collagen [7–10] and have been shown to extravasate into tissues and participate with resident mesenchymal cells in the reparative/fibrogenic process. Although it seems that, regardless the cellular origin, activated myofibroblasts behave similarly as key effectors of fibrogenesis, the possible participation of blood-borne cells and of cells derived from epithelial- or endothelial-mesenchymal transition has raised the possibility of acting therapeutically on their development, activation and recruitment.

Stellate Cells

In organs such as the liver, pancreas and intestine, stellate cells are well established cellular sources of activated myofibroblasts. Stellate cells are characterized by the ability to store retinyl esters in intracytoplasmic lipid droplets and by ultrastructural features of vascular pericytes (i.e. the presence of massive 5-nm actin-like filaments) and they may contribute to reinforce the endothelial lining and/or enhance the efficiency of contraction of capillaries and particularly those with sinusoidal structure and function [11]. The role of vitamin A-storing cells is maximally evident and understandable in the liver, which is a fundamental organ in retinoid metabolism and storage. Regardless, this storage feature is present in stellate cells in other organs and tissues, including the pancreas, lung, kidney, intestine, spleen, adrenal gland, ductus deferens and vocal cords. Hepatic and extrahepatic stellate (HSC) cells form what has been defined 'the stellate cell system', whose embryologic origin is still debated. Because of their morphological similarity, positivity for desmin, vimentin and α-smooth muscle actin, they have been considered of mesenchymal origin for many years. However, when HSC were found to contain a host of neural marker proteins it was speculated that HSC could be of neuro-ectodermal origin [12]. Other studies have suggested the possibility that stellate cells may derive from a common endodermal precursor [13]. Finally, recent studies in humans and in animal models, suggest that HSC may derive from bone marrow precursors [10, 14, 15].

The process of hepatic and pancreatic stellate cell activation and phenotypical transformation into myofibroblasts, as well as the their pro-fibrogenic role have been extensively clarified and represent important basis for the understanding of the fibrogenic process in these organs. A first important element concerns the disruption of the normal ECM pattern that follows tissue injury and acute inflammation. A perturbation in the composition of the normal ECM and/or of the cell-cell relationship between epithelial and mesenchymal cells could also be considered a potent stimulus for the activation and proliferation of stellate cells [16–18].

Regulation of Myofibroblast Recruitment, Proliferation, Survival and Pro-Fibrogenic Activity

Platelet aggregation and activation of the coagulation cascade are the first events following tissue damage and provide the first burst for the wound-healing reaction. Platelets are the first cells recruited to site of injury, as they limit blood loss by forming aggregates at the end of damaged blood vessels and act as a platform for the formation of fibrin from fibrinogen. Furthermore their α-granules are rich in growth factors such as platelet-derived growth factor, transforming growth factor- β (TGF-β) and vascular endothelial growth factor (VEGF) which are released upon activation and are potent stimulators of fibroblast and other mesenchymal cells relevant for tissue healing [3]. This first step in tissue repair leads to the recruitment of inflammatory cells in order to neutralize possible infectious agents and to remove the necrotic tissue [1, 19]. In this phase of the process, local fibroblasts and myofibroblasts are recruited at the site of injury in order to synthesize and secrete ECM components under the control of soluble factors secreted by the cell of the inflammatory infiltrate. It is important to stress that exposure to these mediators, generically defined as 'inflammatory', may be time-limited or chronically present according to the nature, extent and reiteration of parenchymal damage. The same consideration applies to thrombin and other components of the coagulation and complement cascade, whose chronic activation represents a potent pro-fibrogenic stimulus [20]. The steps involved in the organization and functions of the inflammatory infiltrate are regulated by chemokines. Chemokines are leukocyte chemoattractants that cooperate with fibrogenic cytokines in the wound healing reaction and in the development of fibrosis by recruiting myofibroblasts, macrophages and other key effector cells to sites of tissue injury. Although a large number of chemokine signalling pathways are involved in the mechanism of fibrogenesis, the CC- and CXC-chemokine receptor families have consistently exhibited important regulatory roles. In particular, CCL3 (macrophage inflammatory protein 1α) and CC-chemokines such as CCL2 (monocyte chemoattractant protein-1), which are chemotactic for mononuclear phagocytes, were identified as fibrogenic mediators [1, 19].

A major advancement in the biology of activated myofibroblasts derives from the elucidation of the pro-fibrogenic role of a tissue-specific renin-angiotensin-aldosterone system that regulates the local synthesis of angiotensin II. In condition of chronic wound healing angiotensin II is produced locally by activated macrophages and myofibroblasts. In myofibroblasts, angiotensin II stimulates its own production, thereby establishing an autocrine cycle of myofibroblast differentiation and activation. angiotensin II, which has been shown to play an important role in the development hepatic fibrosis [21], exerts its effects by directly inducing NADPH oxidase activity, stimulating TGF-β1 production and triggering fibroblast proliferation and differentiation into collagen-secreting myofibroblasts. In addition, recent data obtained in hepatic stellate cells suggest that angiotensin II, acting in an autocrine fashion, induces phosphorylation of RelA via IKK and the stimulation of NF-κB-dependent transcription of cell survival genes [22, 23], thus contributing to the resistance to apoptotic stimuli observed in chronically activated liver myofibroblasts [24].

Role of Innate and Adaptative Immune Mechanisms

Most fibrotic disorders affecting the GI tract have an infectious aetiology, with bacteria, viruses and multicellular parasites driving chronic tissue damage and inflammation. In addition, it is well established that bacteria contribute to the development of chronic disorders due to altered immune regulation such as inflammatory bowel disease (IBD) [25]. It is becoming increasingly clear that conserved pathogen-associated molecular patterns (PAMPs) found on these organisms contribute to myofibroblast activation [26]. PAMPs are pathogen byproducts, such as lipoproteins, bacterial DNA and double-stranded RNA, which are recognized by pattern recognition receptors (PRRs) present on a wide variety of cells, including fibroblasts [27]. The interaction between PAMPs and PRRs serves as a first line of defence during infection and activates numerous proinflammatory cytokine and chemokine responses. In this context, it is particularly relevant that fibroblasts, myofibroblasts and vascular pericytes express a variety of PRRs, including Toll-like receptors (TLRs), and that their ligands can directly activate these cell types and promote their differentiation into collagen-producing myofibroblasts [26, 28, 29]. In addition, upon stimulation with the TLR4 ligand lipopolysaccharide or the TLR2 ligand lipoteichoic acid, fibroblasts activate MAPK, translocate NFkB and secrete substantial amounts of pro-inflammatory cytokines and chemokines [28]. The interaction between PAMPs and PPRs, particularly TLRs, is also important for the establishment of a pro-inflammatory/pro-fibrogenic condition in a defined vascular district (i.e. the portal circulation), with activation of hepatic stellate cells expressing TLRs by an excessive amount of PAMPs reaching the liver as a consequence of abnormal intestinal permeability in chronic alcohol abuse, diabetes and obesity [30–32].

Abundant lymphocytic infiltration is a hallmark of chronic fibrogenic disorders of the GI tract. Lymphocytes are mobilized to sites of injury and become activated following contact with various antigens, which stimulate the production of lymphokines that further activate macrophages and other local inflammatory cells. Thus, there is significant activation of the adaptive immune response in these diseases. Although inflammation typically precedes the development of fibrosis, several lines of evidence suggest that fibrosis is not always characterized by persistent inflammation, implying that the mechanisms regulating fibrosis are to a certain extent distinct from those controlling inflammation.

It is increasingly evident that development of fibrosis following chronic tissue damage is linked with the development of a CD4+ Th2 cell response (involving IL-4, IL-13, IL-5 and IL-21) [33–37], while potent anti-fibrotic activities for the Th1-associated cytokines IFN-γ and IL-12 have been extensively documented in experimental models of fibrosis, and particularly liver fibrosis [33]. Accordingly, several genes known to be involved in the mechanisms of wound healing and fibrosis were up-regulated in animals exhibiting Th2-polarized inflammation [38, 39]. These include pro-collagens I, III and VI, arginase-1 [40], lysyl oxidase [41, 42], matrix metalloproteinase-2 (MMP-2) [43, 44], MMP-9 [45, 46] and tissue inhibitor of matrix metalloproteinase-1 (TIMP-1) [47, 48]. The Th2 cytokines IL-4, IL-5, IL-13 and IL-21 each have distinct roles in the regulation of tissue remodelling and fibrosis. In particular, IL-4 is considered a potent pro-fibrotic mediator with effects nearly twice as powerful as TGF-β [49]. Receptors for IL-4 are found on many mouse [50] and human fibroblast subtypes [51], and in vitro studies showed the synthesis of the extracellular matrix proteins, types I and III collagen and fibronectin, following IL-4 stimulation. IL-13 shares many functional activities with IL-4 because both cytokines exploit the same IL-4Rα/Stat6 signalling pathways [52].

Role of Oxidative Stress

Involvement of oxidative stress has been documented in all fibrogenic disorders characterized by chronic tissue damage as well as in the relative animal models [for review see

53]. Oxidative stress resulting from increased generation of reactive oxygen intermediates and reactive aldehydes, particularly 4-hydroxynonenal, as well as by decreased efficiency of antioxidant defences, does not represent simply a potentially toxic consequence of chronic tissue injury but actively contributes to excessive tissue remodelling and fibrogenesis. Oxidative stress-related mediators released by damaged or activated neighbouring cells can directly affect the behaviour of myofibroblasts: Reactive oxygen species or the reactive aldehyde HNE have been reported to up-regulate expression of critical genes related to fibrogenesis and inflammation, including procollagen type I, monocyte chemoattractant protein-1 and TIMP-1, possibly through activation of a number of critical signal transduction pathways and transcription factors, including activation of JNKs, AP-1 and NF-kB [53]. In addition to this generic pro-fibrogenic role typical of any condition characterized by chronic tissue damage, oxidative stress represents a predominant pro-fibrogenic mechanism in conditions such as chronic alcoholic hepatitis/pancreatitis and non-alcoholic steatohepatitis. In these settings, perisinusoidal fibrosis may develop independently of evident tissue necrosis and inflammation due to the direct pro-fibrogenic action of reactive oxygen intermediates and reactive aldehydes, including acetaldehyde in the case of chronic alcohol abuse. Interestingly, chronic alcohol abuse may induce fibrosis of duodenal villi which is associated with a transformation of villus juxta-parenchymal cells into active subepithelial myofibroblast-like cells able to produce different ECM components [54].

Fibrogenesis and Angiogenesis are Intimately Connected

Pathological angiogenesis, irrespective of the aetiology, has been extensively described in disorders characterized by an extensive and prolonged necro-inflammatory and fibrogenic process, including disease of the liver, pancreas and intestine. Among fibrogenic disorders affecting the GI tract, the impact of angiogenesis on disease progression is becoming central in chronic liver diseases. In the liver, the formation of new vessels, which is closely associated with the pattern of fibrosis development typical of the different chronic liver diseases (CLDs) [17], leads to the progressive formation of the abnormal angio-architecture distinctive of cirrhosis, i.e. the common end-point of fibrogenic CLDs. From a mechanistic point of view, angiogenesis in fibrogenic disorders can be interpreted according 2 main pathways. First, the process of chronic wound healing is characterized by an over-expression of several growth factors, cytokines and MMPs with an inherent pro-angiogenic action [55]. In particular, platelet-derived growth factor, TGF-Tβ1, fibroblast growth factor and VEGF have been shown to exert a potent pro-fibrogenic and pro-angiogenic role. In addition, an increased gene expression of integrins, β-catenin, ephrins and other adhesion molecules involved in ECM remodelling and angiogenesis has been clearly demonstrated in CLDs [56, 57]. Second, neo-angiogenesis is stimulated in hepatic tissue by the progressive increase of tissue hypoxia due to the progressive capillarization of sinusoids and the consequent impairment of oxygen diffusion from the sinusoids to hepatocytes [58–60]. In this context, it is relevant that activated hepatic stellate cells and other ECM-producing cells such as portal fibroblasts and myofibroblasts produce pro-angiogenic factors, including VEGF and angiopoietin I [61–63]. Moreover, exposure to hypoxia results in up-regulation of VEGF receptors type I (Flt-1) and type II (Flk-1) as well as of Tie-2 (i.e. the receptor for angiopoietin I) in the same cell types [60, 61, 64]. Hypoxia-dependent up-regulation and release of VEGF by human hepatic stellate cells can stimulate, in a paracrine and/or autocrine manner, their non-oriented migration and chemotaxis [64]. The role of bone-marrow derived endothelial precursors (vasculogenesis) in hepatic angiogenesis has been suggested by studies employing animal models of hepatic fibrogenesis [65] and needs to be substantiated in human CLDs.

Reversibility of Fibrosis

In most chronic inflammatory diseases, and particularly those affecting the GI tract, repair cannot be accomplished solely by the regeneration of parenchymal cells, even in tissues where significant regeneration is possible, such as the liver. As already introduced and as illustrated in figure 1, fibrosis then represents the best available solution to maintain tissue continuity and avoid parenchymal collapse. It is controversial whether advanced fibrosis can be reversed to the extent that normal tissue architecture is restored completely. Indeed, there is substantial evidence that, if fibrosis is sufficiently advanced, reversal is no longer possible. Indeed, fibrotic deposition related to recent disease and characterized by the presence of thin reticulin fibres, often in the presence of a diffuse inflammatory infiltrate, is likely fully reversible, whereas long-standing fibrosis – indicated by extensive collagen cross-linking by tissue transglutaminase, presence of elastin, dense acellular/paucicellular ECM and decreased expression and/or activity of specific metalloproteinases – is not [66–68]. Because advanced fibrosis is often hypocellular, it has been suggested that incomplete ECM degradation (irreversible fibrosis) develops when the appropriate cellular mediators (the source of MMPs) are no longer present [68]. Thus, ongoing inflammation might be required for the successful resolution of fibrotic disease [69]. Not surprisingly, the source and identity of key MMPs that mediate the resolution of fibrosis are being intensively investigated. Studies performed in models of liver fibrosis have demonstrated that macrophage depletion at the onset of fibrosis resolution could retard ECM degradation and the loss of activated HSCs [70]. This suggests that macrophages are essential for initiating ECM degradation, perhaps by producing MMPs.

An additional factor limiting the regression of established fibrosis is the already mentioned increased survival of activated myofibroblasts. Increased expression of anti-apoptotic pathways is a hallmark of chronic myofibroblast activation and, for example, expression of the anti-apoptotic protein Bcl-2 is markedly evident in myofibroblast-like cells present in areas of fibrosis in liver tissue obtained from patients with HCV-related cirrhosis [24]. It is therefore plausible that long-term fibrogenesis is characterized, in addition to the biochemical evolution of scar tissue and the lack of an appropriate degradation machinery, by the immovability of a critical mass of pro-fibrogenic cells.

Liver Fibrosis

Progressive accumulation of fibrillar ECM associated with major angioarchitectural changes occurs in the liver generally as a consequence of reiterated liver tissue damage caused by infection [hepatitis B virus(HBV) and hepatitis C virus (HCV)], toxins or drugs (mainly alcohol), metabolism (non-alcoholic fatty liver disease) and autoimmune activity, and the related chronic activation of the wound-healing reaction. The process may result in clinically evident liver cirrhosis and hepatic failure. Cirrhosis is defined as an advanced stage of fibrosis, characterized by the formation of regenerative nodules of liver parenchyma that are separated by and encapsulated in fibrotic septa. In general, in those CLDs evolving towards cirrhosis, a significant accumulation of fibrillar ECM is observed only after a clinical course lasting several years and even decades. For example, in the large majority of patients with chronic hepatitis C there is a long latency period (10–15 years) between HCV infection and the detection of minimal stages of fibrosis, in the presence of an evident and consistent degree of necro-inflammatory activity.

There are, however, at least 2 clinical entities characterized by a fast progression of fibrosis, often referred to as 'fulminant'. One is observed in children affected by bilary athresia or progressive familiar intrahepatic cholestasis, and another, more commonly observed, occurs in a subset of patients who have undergone liver transplantation for HBV- or HCV-related end-stage cirrhosis. In these cases, the time interval between re-infection of the transplanted liver and end-stage disease can be as short as 2–3 years [71]. Although cirrhosis is the common result of progressive fibrogenesis, there are distinct patterns of fibrotic development, related to the underlying disorders causing the fibrosis. Biliary fibrosis, due to the co-proliferation of reactive bile ductules and periductular myofibroblast-like cells at the portal-parenchymal interface, tends to follow a portal-to-portal direction. In contrast, the chronic viral hepatitis pattern of fibrosis is considered the result of portal-central (vein) bridging necrosis, thus originating portal-central septa bridging. Finally, a peculiar type of fibrosis development is observed in alcoholic and metabolic liver diseases (e.g. non-alcoholic steatohepatitis), in which the deposition of fibrillar matrix is concentrated around the sinusoids (capillarization) and around groups of hepatocytes (chicken-wire pattern) [17].

It is now clear that several types of ECM-producing cells contribute to liver fibrosis; however, most of the knowledge on the mechanisms of hepatic fibrogenesis derives from studies performed in the past 20 years on hepatic stellate cells isolated from rodent or human liver [for review see 18]. This knowledge has originated research on the cellular mechanisms of fibrogenesis in other organs of the GI tract and particularly the pancreas [72, 73], and has brought the fibrogenic evolution of CLDs to the attention of clinicians. At the time of writing, the clinical evaluation of disease progression in terms of fibrogenic evolution is one of the hot topics in hepatology.

Pancreatic Fibrosis

The development of irregular tissue fibrosis is a hallmark of chronic pancreatitis and follows the destruction of pancreatic parenchyma and inflammatory cell infiltration, and is accompanied by progressively insufficient pancreatic exocrine and endocrine function. Approximately 70% of chronic pancreatitis cases are caused by alcohol abuse, and the remaining cases are associated with genetic disorders, pancreatic duct obstruction, recurrent acute pancreatitis, autoimmune pancreatitis or unknown mechanisms. The initial event that induces fibrogenesis in the pancreas is an injury that may involve the interstitial mesenchymal cells, duct cells, and/or acinar cells. Damage occurring in any of these tissue compartments is associated with cytokine triggered transformation of resident fibroblasts/pancreatic stellate cells into myofibroblasts and the subsequent production and deposition of ECM. As is the case in other forms of fibrotic disease in the GI tract, the participation of myofibroblasts derived from epithelial-mesenchymal transition and of circulating fibrocytes has been also proposed [74,75]. The fibrogenic development depends on the site of injury and the involved tissue compartment. Deposition of excessive extracellular matrix is predominantly inter(peri)lobular (as in alcoholic chronic pancreatitis), periductal (as in hereditary pancreatitis), periductal and interlobular (as in autoimmune pancreatitis), or diffuse inter- and intralobular (as in obstructive chronic pancreatitis).

In many ways, the development of pancreatic fibrosis recalls the different models of progressive scarring observed in liver tissue following chronic parenchymal damage or bile duct obstruction. Accordingly, it is likely that the 2 basic profibrogenic mechanisms known to be involved in hepatic scarring are also involved in pancreatic fibrogenesis: (1) chronic activation of the wound-healing process with persistent chronic inflammation and progressive substitution of the parenchyma with fibrillar extracellular matrix according to the so-called 'necrosis-fibrosis' sequence, and (2) direct profibrogenic and proinflammatory effects of reactive oxygen species and oxidative stress end products, particularly in alcoholic pancreatitis. However, the main difference between liver and pancreas fibrosis is due to the limited regenerative potential of pancreatic tissue and to its prevalent enzymatic content that causes significant fluid extravasation and tissue oedema. In this direction, it has recently been reported that activated pancreatic stellate cells express the protease activated receptor 2 which interacts with trypsin and tryptase, 2 key pancreatic enzymes involved in the pathogenesis of chronic pancreatitis [76]. Trypsin and tryptase were able to induce stellate cell proliferation and collagen synthesis through activation of c-Jun N-terminal kinase and p38 mitogen activated protein kinase. In addition, pancreatic tissue is more sensitive than liver tissue to abnormal pressure developing within the ductal system, and indeed hypertension within the pancreatic ductal system has been shown to represent a key pro-fibrogenic stimulus inducing pancreatic stellate cell activation [77].

Intestinal Fibrosis

An excessive accumulation of scar tissue in the intestinal wall is a common complication of both forms of IBD (i.e. ulcerative colitis and Crohn's disease). In Crohn's disease, fibrosis can involve the whole thickness of the bowel wall and can cause formation of strictures. Some degree of mild to moderate fibrosis is probably an ordinary event in IBD and it is not associated with evident clinical complications, although it is likely to affect different functions such as adsorption, secretion and control of intestinal permeability. Extensive fibrosis is observed in up to 30% of patients with Crohn's disease with the development of a stricturing or penetrating disease phenotype over a 10-year period [78, 79]. Therefore, from a practical point of view, intestinal fibrosis observed in IBD is mainly characterized by mechanical consequences and, differently from hepatic and pancreatic fibrosis, completely lacks clinical markers indicative of progressive impairment of organ function. In this context, it is relevant that in spite of major therapeutic advances in the treatment of Crohn's disease, the incidence of stricture formation has not markedly changed [80], implying that the progression of intestinal fibrosis in this clinical setting may be at least in part independent from the control of the inflammatory process [80].

Intestinal fibrosis in the context of IBD is traditionally viewed as a slow, unidirectional process, in which inflammation encourages local fibroblasts to multiply and deposit collagen as part of the chronic wound healing reaction. This view, although not necessarily incorrect, is rather simplistic and does not explain why some patients develop transmural fibrosis, strictures, adhesions and perforations while others show only a minor excess in ECM deposition. Overall, the available evidence suggests that all the described general pro-fibrogenic mechanisms are operative in the establishment and progression of intestinal fibrosis and that several cell types acts as effectors, including myofibroblasts derived from epithelial-mesenchymal transition, fibrocytes and intestinal stellate cells [81]. Regardless, the inflammatory process typical of IBD is characterized by an extreme complexity and, more than in other fibrogenic disorders, chronic inflammation is responsible for both tissue damage and repair. In this milieu, the same cell effectors of fibrogenesis (i.e. activated myofibroblasts of different origin) are not likely to be merely effectors of ECM deposition but also relevant players in the modulation of immune responses elicited by the interaction with the enteric commensal microbiota [82, 83].

Clinical Evaluation of Fibrosis Progression/Regression

The assessment of the fibrogenic evolution of chronic diseases affecting the GI tract still mainly relies on the histopathological evaluation of bioptic tissue obtained by percutaneous, laparoscopic or surgical biopsy. In the case of CLDs, the use of liver biopsy has represented and still represents the best standard and is a routine procedure for monitoring disease progression and regression. Accordingly, several dedicated semi-quantitative scoring systems for different CLDs have been developed and are currently employed. Effective non-invasive methodologies for the evaluation of hepatic fibrosis progression and possibly regression following treatment are currently being exploited and validated. These include serum markers, transient elastography and improved imaging techniques or algorithms, including easily available biochemical parameters [for review see 84].

Percutaneous pancreatic biopsy, although technically feasible, is seldom performed and the diagnosis of chronic pancreatic (CP) relies on relevant symptoms, imaging modalities to assess pancreatic structure and assessment of pancreatic function. On the other hand, because the primary lesions of early stage CP are usually focal, fine-needle biopsy examinations may yield false-negative results and, in the absence of definite signs of CP, it is often difficult to differentiate early stage disease from recurrent acute pancreatitis [85]. In addition, the correlation between structural and functional impairment of the pancreas in CP is often poor: patients with severe exocrine insufficiency may have a largely normal pancreatic structure and vice versa [86]. While advanced stages of CP may be diagnosed easily by imaging procedures, the diagnosis of early disease still presents a considerable challenge.

The occurrence of severe intestinal fibrosis in IBD becomes clinically evident with the development of complications, although it may be suspected at physical examination in patients with a thin abdominal wall. The identification of patients who have a high risk of intestinal fibrosis seems to be a realistic goal, as exemplified by genetic studies that have revealed an association of fibrostenotic Crohn's disease with mutations in noD2 [87]. Other profibrotic genotypes probably exist and await identification by genome-wide screening in selected populations of IBD patients. These biomarkers could potentially be used in association with novel imaging techniques that identify early fibrotic changes in the intestinal wall [88].

Potential Anti-Fibrogenic Strategies

The considerable advance in the identification of pro-fibrogenic cell types and the elucidation of several pro-fibrogenic mechanisms has led to a major focus of anti-fibrotic research. Indeed, the well-described pathways of myofibroblast activation, subsequent fibrogenesis, with the potential for apoptosis and reversibility, provides a logical framework to define sites of intervention. Within the GI tract, the search for effective anti-fibrogenic strategies is based on the knowledge gained in the area of stellate cell biology, including the biology of the factors (growth factors, cytokines, etc.) conditioning their pro-fibrogenic attitude [18, 89]. Although this major progress in understanding is fairly recent and, hence, still difficult to be translated into practical strategies, more and more articles published in top specialized journals report on the potent anti-fibrogenic action of old and new drugs, including single agents or mixtures derived from traditional herbal medicine. As with any treatment aimed at curing a chronic disease, any potential anti-fibrotic agent should fulfil 2 main criteria: (1) the treatment should be well tolerated, as it will be provided in multiple administrations over a long period, and (2) the active moiety of the drug should reach a sufficient concentration within the liver, possibly with some cell-specific targeting. There is a growing list of novel mediators and pathways that could be exploited in the development of anti-fibrotic drugs. To name just a few options: cytokine, chemokine and TLR antagonists, angiogenesis inhibitors, anti-hypertensive drugs, TGF-β signalling modifiers, B cell-depleting antibodies and stem/progenitor cell transplantation strategies. As there are many potential targets and strategies, what we need now is a well thought-out plan for translating the available experimental information into clinically effective drugs.

However, there are roadblocks ahead that must be overcome before any treatment can reach the clinic. The most difficult obstacle will be to design effective clinical trials with well-defined clinical endpoints. Therefore, the demand for anti-fibrotic drugs that are both safe and effective is great and will likely continue to increase in the coming years. Current approaches aimed at treating fibrosis are primarily directed at inhibiting cytokines (TGF-β1, IL-13), chemokines, specific MMPs, integrins and angiogenic factors, such as VEGF [19, 89]. Although many of these treatments could prove highly successful, ideally, the best therapy would lead to the complete restoration of the damaged tissue, or at least restore homeostasis to the areas that drive the fibrotic response [90, 91]. Cell based therapies using adult bone marrow-derived progenitor/stem cell technologies might also prove highly successful for the treatment of fibrosis [92].

References

1 Wynn TA: Common and unique mechanisms regulate fibrosis in various fibroproliferative diseases. J Clin Invest 2007;117:524–529.

2 Tomasek JJ, Gabbiani G, Hinz B, Chaponnier C, Brown RA: Myofibroblasts and mechano-regulation of connective tissue remodelling. Nat Rev Mol Cell Biol 2002;3:349–363.

3 Kumar V, Abbas AK, Fausto N: Tissue renewal and repair: regeneration, healing, and fibrosis; in Kumar V, Abbas AK, Fausto N (eds): Pathologic Basis of Disease. Philadelphia, Elsevier Saunders, 2005, pp 87–118.

4 Kalluri R, Neilson EG: Epithelial-mesenchymal transition and its implications for fibrosis. J Clin Invest 2003;112:1776–1784.

5 Willis BC, du Bois RM, Borok Z: Epithelial origin of myofibroblasts during fibrosis in the lung. Proc Am Thorac Soc 2006;3:377–382.

6 Zeisberg EM, Tarnavski O, Zeisberg M, Dorfman AL, McMullen JR, Gustafsson E, et al: Endothelial-to-mesenchymal transition contributes to cardiac fibrosis. Nat Med 2007;13:952–961.

7 Bucala R, Spiegel LA, Chesney J, Hogan M, Cerami A: Circulating fibrocytes define a new leukocyte subpopulation that mediates tissue repair. Mol Med 1994;1:71–81.

8 Quan TE, Cowper SE, Bucala R: The role of circulating fibrocytes in fibrosis. Curr Rheumatol Rep 2006;8:145–150.

9 Brittan M, Hunt T, Jeffery R, Poulsom R, Forbes SJ, Hodivala-Dilke K, et al: Bone marrow derivation of pericryptal myofibroblasts in the mouse and human small intestine and colon. Gut 2002;50:752–757.

10 Russo FP, Alison MR, Bigger BW, Amofah E, Florou A, Amin F, et al: The bone marrow functionally contributes to liver fibrosis. Gastroenterology 2006;130:1807–1821.

11 Wake K: Perisinusoidal stellate cells (fat-storing cells, interstitial cells, lipocytes), their related structure in and around the liver sinusoids, and vitamin A-storing cells in extrahepatic organs. Int Rev Cytol 1980;66:303–353.

12 Geerts A: On the origin of stellate cells: mesodermal, endodermal or neuro-ectodermal? J Hepatol 2004;40:331–334.

13 Kiassov AP, Van Eyken P, van Pelt JF, Depla E, Fevery J, Desmet VJ, Yap SH: Desmin expressing nonhematopoietic liver cells during rat liver development: an immunohistochemical and morphometric study. Differentiation 1995;59:253–258.

14 Baba S, Fujii H, Hirose T, Yasuchika K, Azuma H, Hoppo T, et al: Commitment of bone marrow cells to hepatic stellate cells in mouse. J Hepatol 2004;40:255–260.

15 Suskind DL, Muench MO: Searching for common stem cells of the hepatic and hematopoietic systems in the human fetal liver: CD34+ cytokeratin 7/8+ cells express markers for stellate cells. J Hepatol 2004;40:261–268.

16 Schuppan D, Ruehl M, Somasundaram R, Hahn EG: Matrix as a modulator of hepatic fibrogenesis. Semin Liver Dis 2001;21:351–372.

17 Pinzani M, Rombouts K: Liver fibrosis: from the bench to clinical targets. Dig Liver Dis 2004;36:231–242.

18 Friedman SL: Mechanisms of hepatic fibrogenesis. Gastroenterology 2008;134:1655–1669.

19 Wynn TA: Cellular and molecular mechanisms of fibrosis. J Pathol 2008;214:199–210.

20 Calvaruso V, Maimone S, Gatt A, Tuddenham E, Thursz M, Pinzani M, Burroughs AK: Coagulation and fibrosis in chronic liver disease. Gut 2008;57:1722–1727.

21 Bataller R, Schwabe RF, Choi YH, Yang L, Paik YH, Lindquist J, et al: NADPH oxidase signal transduces angiotensin II in hepatic stellate cells and is critical in hepatic fibrosis. J Clin Invest 2003;112:1383–1394.

22 Oakley F, Teoh V, Ching-A-Sue G, Bataller R, Colmenero J, Jonsson JR, Eliopoulos AG, Watson MR, Manas D, Mann DA: Angiotensin II activates IkappaB kinase phosphorylation of RelA at Ser(536) to promote myofibroblast survival and liver fibrosis. Gastroenterology 2009;136:2334–2344.

23 Pinzani M: Unraveling the spider web of hepatic stellate cell apoptosis. Gastroenterology 2009;136:2061–2063.

24 Novo E, Marra F, Zamara E, Valfre di Bonzo L, Monitillo L, Cannito S, Petrai I, Mazzocca A, Bonacchi A, De Franco RS, Colombatto S, Autelli R, Pinzani M, Parola M: Overexpression of Bcl-2 by activated human hepatic stellate cells: resistance to apoptosis as a mechanism of progressive hepatic fibrogenesis in humans. Gut 2006;55:1174–1182.

25 Arseneau KO, Tamagawa H, Pizarro TT, Cominelli F: Innate and adaptive immune responses related to IBD pathogenesis. Curr Gastroenterol Rep 2007;9:508–512.

26 Meneghin MD, Hogaboam C: Infectious disease, the innate immune response, and fibrosis. J Clin Invest 2007;117:530–538.

27 Akira S, Takeda K: Toll-like receptor signalling. Nat Rev Immunol 2004;4:499–511.

28 Otte JM, Rosenberg IM, Podolsky DK: Intestinal myofibroblasts in innate immune responses of the intestine. Gastroenterology 2003;124:1866–1878.

29 Coelho AL, Hogaboam CM, Kunkel SL: Chemokines provide the sustained inflammatory bridge between innate and acquired immunity. Cytokine Growth Factor Rev 2005;16:553–560.

30 Brun P, Castagliuolo I, Pinzani M, Palu G, Martines D: Exposure to bacterial cell wall products triggers an inflammatory phenotype in hepatic stellate cells. Am J Physiol Gastrointest Liver Physiol 2005;289:G571–G578.

31 Brun P, Castagliuolo I, Di Leo V, Buda A, Pinzani M, Palu' G, Martines D: Increased intestinal permeability in obese mice: new evidence in the pathogenesis of nonalcoholic steatohepatitis. Am J Physiol Gastrointest Liver Physiol 2007;292:G518–G525.

32 Seki E, Brenner DA: Toll-like receptors and adaptor molecules in liver disease: update. Hepatology 2008;48:322–335.

33 Wynn TA, Cheever AW, Jankovic D, Poindexter RW, Caspar P, Lewis FA, et al: An IL-12-based vaccination method for preventing fibrosis induced by schistosome infection. Nature 1995;376:594–596.

34 Cheever AW, Williams ME, Wynn TA, Finkelman FD, Seder RA, Cox TM, et al: Anti-IL-4 treatment of *Schistosoma mansoni* infected mice inhibits development of T cells and non-B, non-T cells expressing Th2 cytokines while decreasing egg-induced hepatic fibrosis. J Immunol 1994;153:753–759.

35 Chiaramonte MG, Donaldson DD, Cheever AW, Wynn TA: An IL-13 inhibitor blocks the development of hepatic fibrosis during a T-helper type 2-dominated inflammatory response. J Clin Invest 1999;104:777–785.

36 Pesce J, Kaviratne M, Ramalingam TR, Thompson RW, Urban JF Jr, Cheever AW, et al: The IL-21 receptor augments Th2 effector function and alternative macrophage activation. J Clin Invest 2006;116:2044–2055.

37 Reiman RM, Thompson RW, Feng CG, Hari D, Knight R, Cheever AW, et al: Interleukin-5 (IL-5) augments the progression of liver fibrosis by regulating IL-13 activity. Infect Immun 2006;74:1471–1479.

38 Hoffmann KF, McCarty TC, Segal DH, Chiaramonte M,Hesse M, Davis EM, et al: Disease fingerprinting with cDNA microarrays reveals distinct gene expression profiles in lethal type 1 and type 2 cytokine-mediated inflammatory reactions. FASEB J 2001;15:2545–2547.

39 Sandler NG, Mentink-Kane MM, Cheever AW, Wynn TA: Global gene expression profiles during acute pathogen-induced pulmonary inflammation reveal divergent roles for Th1 and Th2 responses in tissue repair. J Immunol 2003;171:3655–3667.

40 Hesse M, Modolell M, La Flamme AC, Schito M, Fuentes JM, Cheever AW, et al: Differential regulation of nitric oxide synthase-2 and arginase-1 by type 1/type 2 cytokines in vivo: granulomatous pathology is shaped by the pattern of L-arginine metabolism. J Immunol 2001;167:6533–6544.

41 Decitre M, Gleyzal C, Raccurt M, Peyrol S, Aubert-Foucher E, Csiszar K, et al: Lysyl oxidase-like protein localizes to sites of de novo fibrinogenesis in fibrosis and in the early stromal reaction of ductal breast carcinomas. Lab Invest 1998;78:143–151.

42 Akiri G, Sabo E, Dafni H, Vadasz Z, Kartvelishvily Y, Gan N, et al: Lysyl oxidase-related protein-1 promotes tumor fibrosis and tumor progression in vivo. Cancer Res 2003;63:1657–1666.

43 Wang S, Hirschberg R: BMP7 antagonizes TGFβ-dependent fibrogenesis in mesangial cells. Am J Physiol Renal Physiol 2003;284:F1006–F1013.

44 Cheng S, Lovett DH: Gelatinase A (MMP-2) is necessary and sufficient for renal tubular cell epithelial-mesenchymal transformation. Am J Pathol 2003; 162:1937–1949.

45 Underwood DC, Osborn RR, Bochnowicz S, Webb EF, Rieman DJ, Lee JC, et al: SB 239063, a p38 MAPK inhibitor, reduces neutrophilia, inflammatory cytokines, MMP-9, and fibrosis in lung. Am J Physiol Lung Cell Mol Physiol 2000;279:L895–L902.

46 Heymans S, Lupu F, Terclavers S, Vanwetswinkel B, Herbert JM, Baker A, et al: Loss or inhibition of uPA or MMP-9 attenuates LV remodelling and dysfunction after acute pressure overload in mice. Am J Pathol 2005;166:15–25.

47 Kim H, Oda T, Lopez-Guisa J, Wing D, Edwards DR, Soloway PD, et al: TIMP-1 deficiency does not attenuate interstitial fibrosis in obstructive nephropathy. J Am Soc Nephrol 2001;12:736–748.

48 Vaillant B, Chiaramonte MG, Cheever AW, Soloway PD, Wynn TA: Regulation of hepatic fibrosis and extracellular matrix genes by the Th response: new insight into the role of tissue inhibitors of matrix metalloproteinases. J Immunol 2001;167:7017–7026.

49 Fertin C, Nicolas JF, Gillery P, Kalis B, Banchereau J, Maquart FX: Interleukin-4 stimulates collagen synthesis by normal and scleroderma fibroblasts in dermal equivalents. Cell Mol Biol 1991;37:823–829.

50 Sempowski GD, Beckmann MP, Derdak S, Phipps RP: Subsets of murine lung fibroblasts express membrane-bound and soluble IL-4 receptors: role of IL-4 in enhancing fibroblast proliferation and collagen synthesis. J Immunol 1994;152:3606–3614.

51 Doucet C, Brouty-Boye D, Pottin-Clemenceau C, Canonica GW, Jasmin C, Azzarone B: Interleukin (IL)-4 and IL-13 act on human lung fibroblasts: implication in asthma. J Clin Invest 1998;101:2129–2139.

52 Zurawski SM, Vega F Jr, Huyghe B, Zurawski G: Receptors for interleukin-13 and interleukin-4 are complex and share a novel component that functions in signal transduction. EMBO J 1993;12:2663–2670.

53 Novo E, Parola M: Redox mechanisms in hepatic chronic wound healing and fibrogenesis. Fibrogenesis Tissue Repair 2008;1:5.

54 Casini A, Galli A, Calabrò A, Di Lollo S, Orsini B, Arganini L, Jezequel AM, Surrenti C: Ethanol-induced alterations of matrix network in the duodenal mucosa of chronic alcohol abusers. Virchows Arch 1999;434:127–135.

55 Pinzani M, Marra F: Cytokine receptors and signaling in hepatic stellate cells. Semin Liver Dis 2001; 21:397–416.

56 Shackel NA, McGuinness PH, Abbott CA, Gorrell MD, McCaughan GW: Insights into the pathobiology of hepatitis C virus-associated cirrhosis: analysis of intrahepatic differential gene expression. Am J Pathol 2002;160:641–654.

57 Shackel NA, McGuinness PH, Abbott CA, Gorrell MD, McCaughan GW: Identification of novel molecules and pathogenic pathways in primary biliary cirrhosis: cDNA array analysis of intrahepatic differential gene expression. Gut 2001;49:565–576.

58 Corpechot C, Barbu V, Wendum D, Kinnman N, Rey C, Poupon R, et al: Hypoxia-induced VEGF and collagen I expressions are associated with angiogenesis and fibrogenesis in experimental cirrhosis. Hepatology 2002;35:1010–1021.

59 DeLeve LD: Hepatic microvasculature in liver injury. Semin Liver Dis 2007;27:390–400.

60 Rosmorduc O, Wendum D, Corpechot C, Galy B, Sebbagh N, Raleigh J, et al: Hepatocellular hypoxia-induced vascular endothelial growth factor expression and angiogenesis in experimental biliary cirrhosis. Am J Pathol 1999;155:1065–1073.

61 Ankoma-Sey V, Matli M, Chang KB, Lalazar A, Donner DB, Wong L, et al: Coordinated induction of VEGF receptors in mesenchymal cell types during rat hepatic wound healing. Oncogene 1998;17: 115–121.
62 Aleffi S, Petrai I, Bertolani C, Parola M, Colombatto S, Novo E, et al: Upregulation of proinflammatory and proangiogenic cytokines by leptin in human hepatic stellate cells. Hepatology 2005;42:1339–1348.
63 Wang YQ, Luk JM, Ikeda K, Man K, Chu AC, Kaneda K, et al: Regulatory role of vHL/HIF-1alpha in hypoxia-induced VEGF production in hepatic stellate cells. Biochem Biophys Res Commun 2004;317:358–362.
64 Novo E, Cannito S, Zamara E, Valfre di BL, Caligiuri A, Cravanzola C, et al: Proangiogenic cytokines as hypoxia-dependent factors stimulating migration of human hepatic stellate cells. Am J Pathol 2007; 170:1942–1953.
65 Nakamura T, Torimura T, Sakamoto M, Hashimoto O, Taniguchi E, Inoue K, et al: Significance and therapeutic potential of endothelial progenitor cell transplantation in a cirrhotic liver rat model. Gastroenterology 2007;133:91–107.
66 Ricard-Blum S, Bresson-Hadni S, Vuitton DA, Ville G, Grimaud JA: Hydroxypyridinium collagen crosslinks in human liver fibrosis: study of alveolar echinococcosis. Hepatology 1992;15:599–602.
67 Hayasaka A, Ilda S, Suzuki N, Kondo F, Miyazaki M, Yonemitsu H: Pyridinoline collagen cross-links in patients with chronic viral hepatitis and cirrhosis. J Hepatol 1996;24:692–698.
68 Issa R, Zhou X, Constandinou CM, Fallowfield J, Millward-Sadler H, Gaca MD, Sands E, Suliman I, Trim N, Knorr A, Arthur MJ, Benyon RC, Iredale JP: Spontaneous recovery from micronodular cirrhosis: evidence for incomplete resolution associated with matrix cross-linking. Gastroenterology 2004;126:1795–1808.
69 Iredale JP: Models of liver fibrosis: exploring the dynamic nature of inflammation and repair in a solid organ. J Clin Invest 2007;117:539–548.
70 Duffield JS, Forbes SJ, Constandinou CM, Clay S, Partolina M, Vuthoori S, et al: Selective depletion of macrophages reveals distinct, opposing roles during liver injury and repair. J Clin Invest 2005;115:56–65.
71 Poynard T, Bedossa P, Opolon P: Natural history of liver fibrosis progression in patients with chronic hepatitis C: the OBSVIRC, METAVIR, CLINIVIR, and DOSVIRC groups. Lancet 1997;349:825–832.
72 Pinzani M: New kids on the block: pancreatic stellate cells enter the fibrogenesis world. Gut 1999;44: 451–452.
73 Pinzani M: Pancreatic stellate cells: new kids become mature. Gut 2006;55:12–14.
74 Shimizu K: Mechanisms of pancreatic fibrosis and applications to the treatment of chronic pancreatitis. J Gastroenterol 2008;43:823–832.
75 Vonlaufen A, Wilson JS, Apte MV: Molecular mechanisms of pancreatitis: current opinion. J Gastroenterol Hepatol 2008;23:1339–1348.
76 Masamune A, Kikuta K, Satoh M, et al: Protease-activated receptor-2-mediated proliferation and collagen production of rat pancreatic stellate cells. J Pharmacol Exp Ther 2005;312:651–658.
77 Watanabe S, Nagashio Y, Asaumi H, Nomiyama Y, Taguchi M, Tashiro M, et al: Pressure activates rat pancreatic stellate cells. Am J Physiol Gastrointest Liver Physiol 2004;287:G1175–G1181.
78 Silverstein MD, Loftus EV, Sandborn WJ, Tremaine WJ, Feagan BG, Nietert PJ, Harmsen WS, Zinsmeister AR: Clinical course and costs of care for Crohn's disease: Markov model analysis of a population-based cohort. Gastroenterology 1999; 117:49–57.
79 Louis E, Collard A, Oger AF, Degroote E, Aboul Nasr El Yafi FA, Belaiche J: Behaviour of Crohn's disease according to the Vienna classification: changing pattern over the course of the disease. Gut 2001;49:777–782.
80 Cosnes J, Nion-Larmurier I, Beaugerie L, et al: Impact of the increasing use of immunosuppressants in Crohn's disease on the need for intestinal surgery. Gut 2005;54:237–241.
81 Rieder F, Fiocchi C: Intestinal fibrosis in IBD: a dynamic, multifactorial process. Nat Rev Gastroenterol Hepatol 2009;6:228–235.
82 Sartor RB: Microbial influences in inflammatory bowel diseases. Gastroenterology 2008;134:577–594.
83 Fiocchi C: Immune-nonimmune cell interactions: the other crosstalk between innate and adaptive immunity; in Hibi T (ed): Recent Advances in Inflammatory Bowel Disease. Tokyo, Elsevier, 2007, pp 29–35
84 Pinzani M, Vizzutti F, Arena U, Marra F: Technology insights: non invasive assessment of liver fibrosis by biochemical scores and elastography. Nat Rev Gastroenterol Hepatol 2008;5:95–106.
85 Witt H, Apte MV, Keim V, Wilson JS: Chronic pancreatitis: challenges and advances in pathogenesis, genetics, diagnosis, and therapy. Gastroenterology 2007;132:1557–1573.
86 Bozkurt T, Braun U, Leferink S, Gilly G, Lux G: Comparison of pancreatic morphology and exocrine functional impairment in patients with chronic pancreatitis. Gut 1994;35:1132–1136.

87 Abreu MT, Taylor KD, Lin YC, Hang T, Gaiennie J, Landers CJ, Vasiliauskas EA, Kam LY, Rojany M, Papadakis KA, Rotter JI, Targan SR, Yang H: Mutations in NOD2 are associated with fibrostenosing disease in patients with Crohn's disease. Gastroenterology 2002;123:679–688.
88 Danese S, Sans M, Spencer DM, Beck I, Doñate F, Plunkett ML, de la Motte C, Redline R, Shaw DE, Levine AD, Mazar AP, Fiocchi C: Angiogenesis blockade as a new therapeutic approach to experimental colitis. Gut 2007;56:855–862.
89 Pinzani M, Rombouts K, Colagrande S: Fibrosis in chronic liver diseases: diagnosis and management. J Hepatol 2005;42 (suppl 1):S22–S36.
90 Iredale JP, Benyon RC, Pickering J, McCullen M, Northrop M, Pawley S, et al: Mechanisms of spontaneous resolution of rat liver fibrosis: hepatic stellate cell apoptosis and reduced hepatic expression of metalloproteinase inhibitors. J Clin Invest 1998;102: 538–549.
91 Wright MC, Issa R, Smart DE, Trim N, Murray GI, Primrose JN, et al: Gliotoxin stimulates the apoptosis of human and rat hepatic stellate cells and enhances the resolution of liver fibrosis in rats. Gastroenterology 2001;121:685–698.
92 Caplan AI, Dennis JE: Mesenchymal stem cells as trophic mediators. J Cell Biochem 2006;98:1076–1084.

Massimo Pinzani, MD, PhD, Professor of Medicine
Dipartimento di Medicina Interna, Università degli Studi di Firenze
Viale G.B. Morgagni, 85
IT–50134 Firenze (Italy)
Tel. +39 055 429 6273, Fax +39 055 417 123, E-Mail m.pinzani@dmi.unifi.it

Mayerle J, Tilg H (eds): Clinical Update on Inflammatory Disorders of the Gastrointestinal Tract.
Front Gastrointest Res. Basel, Karger, 2010, vol 26, pp 32–41

Chronic Hepatitis B: Pathophysiology, Diagnosis and Treatment Options

Karsten Wursthorn · Ingmar Mederacke · Michael P. Manns

Department of Gastroenterology, Hepatology and Endocrinology, Hannover Medical School, Hannover, Germany

Abstract

Chronic infection with hepatitis B virus (HBV) today affects approximately 350 million people worldwide and about 2 billion people show serological incidence of prior or current infection. Between 15 and 40% of the chronically infected patients are at risk of developing end-stage liver disease including liver cirrhosis and hepatocellular carcinoma. The method of acquiring HBV infection varies geographically. Perinatal transmission is the most common route in high prevalence areas such as Asia or Africa, and is often clinically asymptomatic. It leads to HBsAg positive, chronic HBV infection in the majority of the cases. Over 95% of acutely infected adolescents and adults resolve the infection spontaneously and do not develop chronic hepatitis B. In chronic HBV infection, 3 distinct phases influenced by the immune system can be identified. HBV is not directly cytopathogenic. Hepatocellular injury is mainly mediated by the host's immune response to viral antigens on the surface of the infected hepatocytes. In acute, self-limiting hepatitis B, strong T cell responses to HBV antigens are detectable. In chronic carriers, virus-specific T cell responses in the peripheral blood are diminished. Antibody responses are strong in both settings. Loss of immunological control in immunosuppressed patients can lead to reactivation of HBV DNA replication and hepatitis. Treatment options include therapies to stimulate the immune system, such as IFN-α, and antiviral therapy in the form of nucleoside/nucleotide reverse transcriptase inhibitors of the HBV polymerase. With the recent approval of potent antiviral drugs, serum HBV DNA reduction often leading to improved clinical outcomes can be achieved in the majority of the cases.

Chronic infection with hepatitis B virus (HBV) today affects approximately 350 million people worldwide, and about 2 billion people show serological incidence of prior or current infection [1]. Between 15 and 40% of chronically infected patients are at risk of developing end-stage liver disease, including liver cirrhosis and hepatocellular carcinoma [2]. This entity accounts for approximately 0.6–1 million or 2.5% of deaths worldwide according to World Health Organization's 2002 estimate.

The method of acquiring HBV infection varies geographically and so does the course of the disease. In high prevalence areas, such as China, South-East Asia and

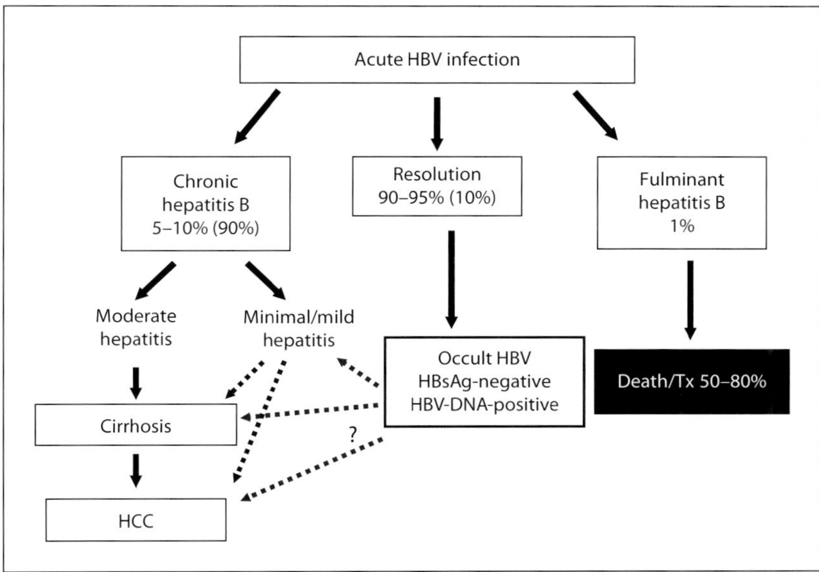

Fig. 1. Course of acute HBV infection. Numbers in parantheses refer to respective percentages in children. HCC = Hepatocellular carcinoma; Tx = medical treatment.

sub-Saharan Africa, where ≥8% of the population are HBsAg positive, perinatal transmission is the most common route. When HBV infection is acquired at birth or during early childhood, most infections are asymptomatic. Perinatal transmission carries the highest risk of developing chronic infection, which occurs in approximately 90% of the cases [3]. This is in contrast to regions with low prevalence, such as the USA, Canada and western Europe where ≤2% of the population are HBsAg positive and sexual contact and percutaneous transmission is predominant. Infection leads to acute HBV-related disease more often in the adolescent and adult population and symptoms include jaundice and nausea. Less than 5% of infections acquired in adulthood become chronic. While the risk of chronic HBV infection decreases with age, the risk of clinically apparent hepatitis rises [4] (fig. 1).

In patients who have an impaired immune system the prevalence of HBsAg is higher. The course of the hepatitis B infection is often severely aggravated. In western Europe and the USA, 6–14% of HIV patients have HBV co-infection, leading to an unfavourable outcome of the liver disease [5]. HBV/HIV co-infection is associated with increased progression of liver fibrosis and cirrhosis, as well as higher rates of liver decompensation, hepatocellular carcinoma and liver-related mortality. HBV reactivation in patients receiving cytotoxic chemotherapy and/or immunosuppressive therapy for haemato-oncological malignancies or autoimmune diseases can lead to recovery, persistence, or liver failure and death [6]. Reactivation of HBV infection is the loss of protective anti-HBs antibodies associated with the reappearance

Table 1. Phases of chronic hepatitis B (after Hadziyannis [31])

Phase of hepatitis B	HBe antigen status	Serum HBV DNA	Serum ALT
Immune tolerant	positive	high	normal
Immune active	positive and/or negative	fluctuating	elevated
Inactive	negative	low	normal and/or elevated
Anti-HBc only	negative	negative	normal
Resolved	negative	negative	normal

of HBsAg. It was first described in patients receiving anti-tumour chemotherapy for myeloproliferative and lymphoproliferative disorders [7].

A third group of patients requiring special attention by healthcare providers is liver transplant recipients. The main goal here is to prevent recurrence of HBV infection in patients receiving transplants for hepatitis B related liver disease. The introduction of hepatitis B immune globulin has lowered the risk for HBV recurrence and the addition of the nucleoside analogue lamivudine further improved the outcome of HBsAg-free survival after transplantation [8].

Phases of Hepatitis B Infection

The first and most important differentiation in HBV infection is between acute and chronic courses. In HBs-antigen positive, chronic HBV infection, 3 distinct phases of HBV infection can be identified, the immune-tolerant phase, the immune-active phase and the inactive HBsAg carrier phase (table 1). Since evidence is mounting that serum HBV DNA level plays a central role in disease progression, classification of chronic hepatitis B might be further simplified by taking into account only high and low viremic HBsAg carrier status [9]. In addition, HBeAg-negative 'inactive' carriers comprise a heterogeneous group of patients, who sometimes have elevated ALT levels and serum HBV DNA pointing to the possibility of ongoing hepatic damage. Therefore, the description of this cohort as 'inactive' might not be correct. An exception to resolved hepatitis B is the 'anti-HBc-only' status, which has no clinical relevance except in a transplantation setting or under immunosuppression [10]. The prevalence of the 'anti-HBc-only status in developed countries is high, especially in older patients and patients with HCV co-infection.

Role of Viral Load

Serum HBV DNA concentrations reflect the level of hepatocellular viral replication. There is a strong correlation between cccDNA levels, levels of total intracellular HBV

DNA and serum HBV DNA [11]. cccDNA can be detected in the liver of HBsAg negative patients with evidence of HBV clearance.

There is no threshold level for serum HBV DNA associated with the progression of liver disease. High viral load ($>10^5$ copies/ml) at baseline has been shown to increase the risk for liver-related mortality and HCC. More recent studies have shown that serum HBV DNA $\leq 1 \times 10^5$ copies/ml can also lead to liver cirrhosis and holds an increased risk for the development of hepatocellular carcinoma [12].

The most comprehensive data so far on HBV DNA viral load and liver disease comes from the REVEAL-HBV study. More than 3,500 Taiwanese patients with chronic HBV infection were recruited to this prospective cohort study in 1991 and 1992 and analyzed for the incidence of liver disease and hepatocellular carcinoma in relation to serum HBV DNA concentration at baseline [13, 14]. For the evaluation of the risk for developing cirrhosis, a cohort of 3,852 HBsAg-positive patients without evidence of co-infection was screened regularly for a mean follow-up of 11 years. There were 365 cases of cirrhosis newly diagnosed. Out of the 1,563 patients with HBV DNA $\geq 10^4$ copies/ml at baseline, 274 patients developed cirrhosis, accounting for 75% of all cases. Increasing HBV DNA level was associated with increasing risk for cirrhosis incidence after adjusting for age, sex, smoking and alcohol use. The relative risk for cirrhosis increased from 1.4 in patients with HBV DNA $<9.9 \times 10^3$ copies/ml to 2.5 in patients with HBV DNA level $\geq 10^4$ copies/ml and was highest in the group of patients with $>10^6$ copies/ml, with a relative risk of 9.8. Other factors correlating with the development of liver cirrhosis were sex, age, HBeAg status and an increased serum ALT level [14]. In a second analysis with the same cohort of 3,653 HBsAg-positive participants aged 30–65 years, 164 cases of hepatocellular carcinoma and 346 deaths occurred during the mean follow-up of 11.4 years. Similar to the risk for liver cirrhosis, incidence rates for HCC increased with higher baseline HBV DNA levels in a dose-response relationship.

Immunopathogenesis

The magnitude and width of the immune response during acute HBV infection is the major determinant for the outcome of infection. While a multi-specific, polyclonal HBV-specific CD4 and CD8 T cell response is associated with elimination of the virus, a weak, oligoclonal immune response is accompanied by viral persistence [15].

For a period of 4–7 weeks after infection, there is no HBV DNA detectable in the blood. But within weeks viral load increases to levels of up to 10^{10} copies/ml and the majority of hepatocytes become infected [16]. The initial phase of viral infections is controlled by components of the innate immunity, including production of type I IFN and TNF-α. There is no activation of intrahepatic cellular genes during the initial phase of infection, when no HBV replication is detectable [17]. This changes dramatically immediately after the exponential replication of HBV begins. Studies on experimentally infected chimpanzees have shown that at this stage a strong HBV-specific IFN-γ and TNF-α response and a wide range of intrahepatic cellular genes are activated [16, 17].

For the elimination of the virus a strong CD4 and CD8 T cell response is necessary as well as a humoral immune response. Depletion experiments in chimpanzees showed that CD8 T cells are the main effector cells for viral clearance and disease pathogenesis [18]. In the peripheral blood of patients with resolved hepatitis B infection CD8+ T cells with strong cytotoxic activity against different epitopes of HBcAg, HBsAg and polymerase are detectable, whereas in patients with chronic HBV infection HBV-specific cytotoxic T cells have a low frequency [19]. Similar to patients with acute, self-limiting HBV-infection, the HLA-class II T cell response against the HBV nucleocapsid (HBcAg) is strong during seroconversion from HBeAg to anti-HBeAg in patients with chronic HBV infection. The CD4 T cell response against the nucleocapsid seems to be a key for the control of the infection.

The importance of the humoral immune response is obvious, since antibodies against HBsAg protect hepatocytes against an infection with HBV and neutralize viral particles in the blood. Anti-HBs antibodies are present in patients with resolved infection but they are missing in patients with chronic HBV infection, where they are probably produced in a lower number and thus are neutralized by the redundant HBsAg. The function of antibodies against HBeAg and HBcAg is unclear. These antibodies are detected during the acute and chronic course in high concentrations, but they do not seem to have a protective effect. However, patients with chronic HBV infection and antibodies against HBeAg usually reached the immune control phase of the infection and have lower levels of HBV DNA. In the setting of liver transplantation, antibodies against HBcAg in the organ recipient or donor indicate a resolved hepatitis B infection and require antiviral therapy and administration of hepatitis B immunoglobulins after transplantation to prevent a reactivation of HBV.

In patients with chronic infection, ongoing HBV replication and production of viral antigens could be a reason for the exhaustion of virus specific adaptive immunity. Other possible causes for the weak T cell response are dendritic cell impairment, regulatory T cells or the immunological features of the liver environment [20].

Both innate and adaptive immunity are pivotal for the control of HBV activation. The different components of the immune system are so intimately connected with each other that the failure of one part can affect the function of another. Lack of CD4+ T cells alters the function of CD8+ T-cells and antibody production. An impaired function of virus-specific CD8+ T-cells cannot be compensated for by antibodies alone.

The cure of an HBV infection is characterized by lifelong immunity, which is mediated by HBs antibodies and HBV-specific CD4 and CD8 T cells. Traces of the virus are still detectable years after resolution, which seems to be important for maintaining sustained immunity [21]. This assumption is supported by the fact that following

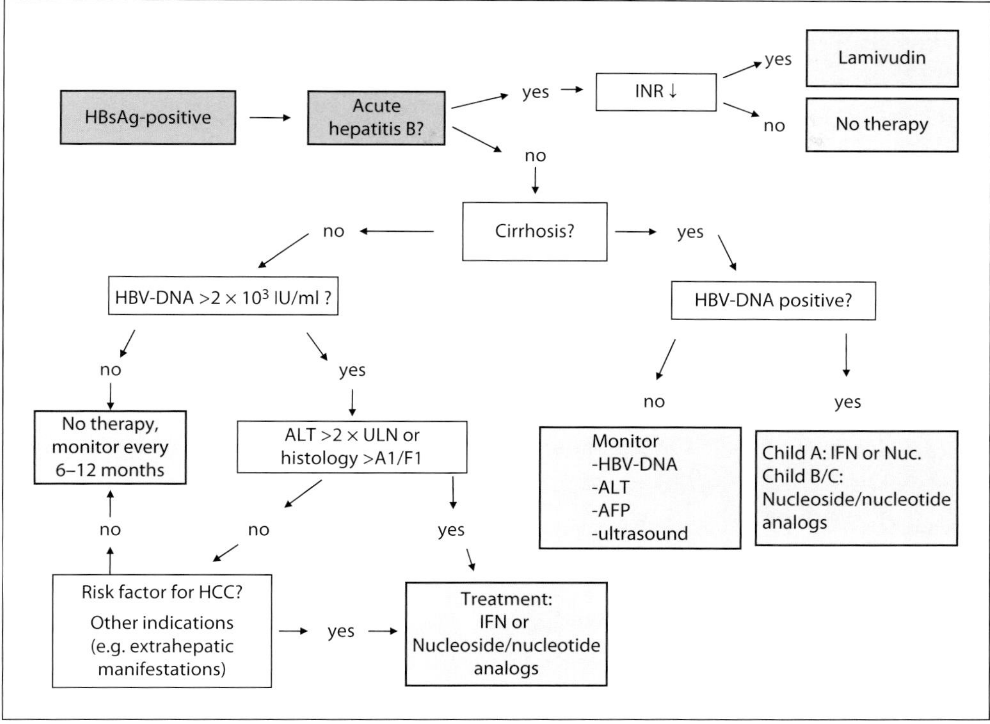

Fig. 2. Treatment algorithms according to the German guidelines for the treatment of hepatitis B. AFP = Alpha feto protein; HCC = hepatocellular carcinoma; Nuc. = nucleoside/nucleotide analogs; ULN = upper limit of normal.

procedures that induce severe immunosuppression, such as chemotherapy or after transplantations, resolved HBV infections can reactivate.

During the course of an HBV infection, complete elimination of the virus is rarely achieved. Instead, there is immunological control of viral replication, as is the case in herpes simplex virus, cytomegalovirus and Epstein-Barr virus.

Treatment Options in Chronic Hepatitis B

Figure 2 gives an overview of the treatment algorithms according to the German guidelines for hepatitis B. Having replaced standard IFN for most indications, pegylated IFN is a very effective agent for the treatment of HBeAg+ chronic hepatitis B with low viral load, genotype A and high serum ALT resulting in a high rate of sustained HBeAg seroconversion, even in patients with advanced fibrosis but compensated liver disease [22] (table 2).

Lamivudine was the first nucleoside analogue to be administered and approved for the treatment of chronic hepatitis B. Lamivudine at a dose of 100 mg once daily

Table 2. Approved drugs for the treatment of chronic hepatitis B

Substance	Brand name	Approved dose
IFN-α		
Pegylated IFN-α 2a	Pegasys®	180 μg, once weekly for 48 weeks
IFN-α 2a	Roferon®	2.5–5 million IU/m² body surface, 3 times/week for 4–6 months
IFN-α 2b	Intron A®	5–10 million IU, 3 times/week (every second day), s.c., for 4–6 months
Nucleoside analogues		
Lamivudine	Zeffix®	100 mg, once dailyy
Entecavir	Baraclude®	0.5 mg, once daily; 1.0 mg for patients with lamivudine resistance
Telbivudine	Sebivo®	600 mg, once daily
Nucleotide analogues		
Adefovir dipivoxil	Hepsera®	10 mg, once daily
Tenofovir disoproxil	Viread®	245 mg, once daily

has shown antiviral activity in patients infected with HBV, with HBV DNA suppression and loss of hepatitis Be antigen as well as biochemical response and histological improvement after 1 year of treatment [23]. However, the development of resistance is a major problem, especially with lamivudine. Long-term administration leads to high rates of resistance in the HBV polymerase gene (up to 70% after 5 years), which compromises consecutive therapies.

The nucleotide analogue adefovir has the poorest antiviral effect for HBV but a lower incidence of resistance [24]. Partial cross-resistance with tenofovir and lamivudine occurs, the latter due to a mutation at rtA181T/V.

Together with tenofovir, entecavir is the most potent drug with regard to antiviral activity against HBV. It leads to undetectable HBV DNA more frequently than lamivudine and adefovir [25] and is effective against HBV that is resistant to these 2 drugs. Entecavir, a deoxyguanosine analogue, by itself has a high genetic barrier with resistance occurring only after the accumulation of multiple changes in the HBV polymerase gene. However, previous lamivudine therapy, increases the risk of developing resistance since crucial mutations at the HBV polymerase gene are more frequent in lamivudine-experienced patients [26]. As a consequence, the recommended dose of entecavir is doubled in these patients (to 1 mg/day).

With telbivudine therapy, in HBeAg+ and in HBeAg– patients, telbivudine resistance mutations are selected for after 2 years in 25.1 of HBeAg+ patients and 10.8% of HBeAg– patients. Typically, selection is for the M204I mutation, which leads to cross-resistance to drugs including lamivudine [27]. The rate of resistance after 2 years is usually lower in those patients with undetectable HBV DNA after 24 weeks of antiviral therapy (4% in the HBeAg+, 2% in the HBeAg– population). Therefore, telbivudine is recommended for the treatment of hepatitis B in patients with lower HBV DNA counts or as an add-on drug in case of adefovir/tenofovir resistance [28]. Telbivudine is listed as a pregnancy category B drug by the US Food and Drug Administration.

Tenofovir is an adenosine nucleotide analogue with strong antiviral potency and a high genetic barrier in chronic HBV infection. It is superior to Adefovir in treatment-naïve, HBeAg+ and HBeAg– mono-infected patients in terms of HBV DNA negativity after 1 year [29]. ALT normalization and loss of hepatitis Bs antigen was more frequent in HBeAg+ patients receiving tenofovir after 1 year compared to the group receiving adefovir [29]. When switched from adefovir to tenofovir in the second year, the rate of patients with undetectable HBV DNA approximated the rate in the tenofovir group. Tenofovir is also potent in HBV mono-infected patients with lamivudine pre-treatment or resistance [30]. Resistance to tenofovir has not been described so far in these patients. Tenofovir is also listed as pregnancy category B drug by the US Food and Drug Administration.

Emtricitabine is a cytosine analogue which is not approved for the treatment of chronic HBV infection but is known to have HBV-directed antiviral properties. Emtricitabine suppresses HBV replication in >50% of patients with HBV mono-infection.

Summary

Viral hepatitis is the main source of liver disease. Progression leading to liver cirrhosis and hepatocellular carcinoma occurs frequently in patients with chronic HBV or HCV infection. There are various factors – viral and host-related, modifiable and non-modifiable – that are associated with advancing liver disease. For HBV, viral factors include HBeAg status, HBV genotype, naturally occurring mutations and viral load. The level of serum HBV DNA reflecting the concentration of intrahepatic HBV DNA is positively correlated with the risk for liver cirrhosis and hepatocellular carcinoma. The continued presence of intracellular HBV and the resulting immune response might be the driving force for the progression of HBV-related liver disease and HBV reactivation. However, little is known about the natural course of intrahepatic HBV DNA and further studies are needed. Effective antiviral treatment with a high genetic barrier and strong antiviral activity is now available, enabling healthcare providers to effectively reduce viral replication and thereby improve the long-term outcome of patients with chronic HBV infection.

References

1 Lee WM: Hepatitis B virus infection. N Engl J Med 1997;337:1733–1745.
2 McMahon BJ: Epidemiology and natural history of hepatitis B. Semin Liver Dis 2005;25:3–8.
3 Stevens CE, Beasley RP, Tsui J, et al: Vertical transmission of hepatitis B antigen in Taiwan. N Engl J Med. 1975;292:771–774.
4 McMahon BJ, Alward WL, Hall DB, et al: Acute hepatitis B virus infection: relation of age to the clinical expression of disease and subsequent development of the carrier state. J Infect Dis 1985;151: 599–603.
5 Alter MJ: Epidemiology of viral hepatitis and HIV co-infection. J Hepatol 2006;44(1 suppl):S6–S9.
6 Xunrong L, Yan AW, Liang R, et al: Hepatitis B virus (HBV) reactivation after cytotoxic or immunosuppressive therapy: pathogenesis and management. Rev Med Virol 2001;11:287–299.
7 Wands JR, Chura CM, Roll FJ, et al: Serial studies of hepatitis-associated antigen and antibody in patients receiving antitumor chemotherapy for myeloproliferative and lymphoproliferative disorders. Gastroenterology 1975;68:105–112.
8 Samuel D, Muller R, Alexander G, et al: Liver transplantation in European patients with the hepatitis B surface antigen. N Engl J Med 1993;329:1842–1847.
9 Cornberg M, Protzer U, Dollinger M, et al: Prophylaxis, diagnosis and therapy of hepatitis B virus (HBV) infection: the German guidelines for the management of HBV infection. Z Gastroenterol 2007;45:1281–1328.
10 Ciesek S, Helfritz FA, Lehmann U, et al: Persistence of occult hepatitis B after removal of the hepatitis B virus-infected liver. J Infect Dis 2008;197:355–360.
11 Werle-Lapostolle B, Bowden S, Locarnini S, et al: Persistence of cccDNA during the natural history of chronic hepatitis B and decline during adefovir dipivoxil therapy. Gastroenterology 2004;126:1750–1758.
12 Tang B, Kruger WD, Chen G, et al: Hepatitis B viremia is associated with increased risk of hepatocellular carcinoma in chronic carriers. J Med Virol 2004;72:35–40.
13 Chen CJ, Yang HI, Su J, et al: Risk of hepatocellular carcinoma across a biological gradient of serum hepatitis B virus DNA level. JAMA 2006;295:65–73.
14 Iloeje UH, Yang HI, Su J, et al: Predicting cirrhosis risk based on the level of circulating hepatitis B viral load. Gastroenterology 2006;130:678–686.
15 Bertoletti A, Ferrari C: Kinetics of the immune response during HBV and HCV infection. Hepatology 2003;38:4–13.
16 Guidotti LG, Rochford R, Chung J, et al: Viral clearance without destruction of infected cells during acute HBV infection. Science 1999;284:825–829.
17 Wieland S, Thimme R, Purcell RH, et al: Genomic analysis of the host response to hepatitis B virus infection. Proc Natl Acad Sci USA 2004;101:6669–6674.
18 Thimme R, Wieland S, Steiger C, et al: CD8(+) T cells mediate viral clearance and disease pathogenesis during acute hepatitis B virus infection. J Virol 2003;77:68–76.
19 Chisari FV: Rous-Whipple Award Lecture: Viruses, immunity, and cancer: lessons from hepatitis B. Am J Pathol 2000;156:1117–1132.
20 Bertoletti A, Gehring AJ: The immune response during hepatitis B virus infection. J Gen Virol 2006; 87:1439–1449.
21 Rehermann B, Ferrari C, Pasquinelli C, et al: The hepatitis B virus persists for decades after patients' recovery from acute viral hepatitis despite active maintenance of a cytotoxic T-lymphocyte response. Nat Med 1996;2:1104–1108.
22 Lau GK, Piratvisuth T, Luo KX, et al: Peginterferon Alfa-2a, lamivudine, and the combination for HBeAg-positive chronic hepatitis B. N Engl J Med 2005;352:2682–2695.
23 Dienstag JL, Schiff ER, Wright TL, et al: Lamivudine as initial treatment for chronic hepatitis B in the United States. N Engl J Med 1999; 341:1256–1263.
24 Marcellin P, Chang TT, Lim SG, et al: Long-term efficacy and safety of adefovir dipivoxil for the treatment of hepatitis Be antigen-positive chronic hepatitis B. Hepatology 2008;48:750–758.
25 Chang TT, Gish RG, de Man R, et al: A comparison of entecavir and lamivudine for HBeAg-positive chronic hepatitis B. N Engl J Med 2006;354:1001–1010.
26 Tenney DJ, Rose RE, Baldick CJ, et al: Two-year assessment of entecavir resistance in lamivudine-refractory hepatitis B virus patients reveals different clinical outcomes depending on the resistance substitutions present. Antimicrob Agents Chemother 2007;51:902–911.
27 Thio CL, Locarnini S: Treatment of HIV/HBV coinfection: clinical and virologic issues. AIDS Rev 2007;9:40–53.
28 European Association for the Study of the Liver: EASL Clinical Practice Guidelines: Management of chronic hepatitis B. J Hepatol 2009;50:227–242.

29 Marcellin P, Heathcote EJ, Buti M, et al: Tenofovir disoproxil fumarate versus adefovir dipivoxil for chronic hepatitis B. N Engl J Med 2008;359:2442–2455.
30 van Bommel F, Wunsche T, Mauss S, et al: Comparison of adefovir and tenofovir in the treatment of lamivudine-resistant hepatitis B virus infection. Hepatology 2004;40:1421–1425.
31 Hadziyannis SJ, Papatheodoridis GV: Hepatitis B e antigen-negative chronic hepatitis B: natural history and treatment. Semin Liver Dis 2006;26:130–141.

Karsten Wursthorn
Carl-Neuberg-Strasse 1
DE–30623 Hannover (Germany)
Tel. +49 511 532 3305, Fax +49 511 532 8142, E-Mail wursthorn.mhh@gmail.com

Mayerle J, Tilg H (eds): Clinical Update on Inflammatory Disorders of the Gastrointestinal Tract.
Front Gastrointest Res. Basel, Karger, 2010, vol 26, pp 42–58

Chronic Hepatitis C: Pathophysiology, Diagnosis and Treatment Options

Tarik Asselah[a] · Vassili Soumelis[b] · Emilie Estrabaud[a] · Patrick Marcellin[a]

[a]Service d'hépatologie and INSERM U773, Hôpital Beaujon AP-HP and Université Denis Diderot Paris 7, and [b]Department of Immunology and INSERM U932, Institut Curie, Paris, France

Abstract

Hepatitis C is among the leading causes of chronic liver disease worldwide, with a prevalence of approximately 170 million cases. The severity of disease varies from asymptomatic chronic infection to cirrhosis and hepatocellular carcinoma. The main treatment goal in chronic hepatitis C is the prevention of cirrhosis and hepatocellular carcinoma by eradicating the virus. In the last decade, advances have been made in treatment with the combination of pegylated interferon and ribavirin. At present, in patients with hepatitis C virus (HCV), therapy results in a sustained virological response in approximately 55% of cases. In patients with HCV genotypes 2 or 3, the response rate is about 80% and in genotype 1 it is 50%. Based on existing results, the sustained virological response with this treatment option appears to be associated with viral eradication and a histological benefit and probably also with a reduction in the risk of cirrhosis and hepatocellular carcinoma. Despite this progress, treatment failure still occurs in about half of the patients. Furthermore, therapy results in several side effects and high costs. Premature withdrawal due to adverse events was required in 10–14% of participants in registration trials. These limitations have led to important development of novel compounds under the name of specifically targeted antiviral therapy for HCV (STAT-C). The development of new molecules such as viral enzyme inhibitors (protease and polymerase) is ongoing. Several protease and polymerase inhibitors are under development. So far, promising results have been reported with 2 protease inhibitors (boceprevir and telaprevir). Also, considering side effects and treatment costs, finding more reliable markers to predict virological non-response before starting therapy is essential. The management of HCV infection must include better knowledge of viral cycle and mechanisms of non-response.

Viral Cycle

Hepatitis C virus (HCV) is a major cause of chronic liver disease, with about 170 million people infected worldwide [1]. HCV, identified in 1989, is an enveloped virus with a 9.6-kb single-stranded RNA genome [2–6]. It is a member of the *Flaviviridae* family, genus *Hepacivirus*. HCV replication is error-prone, which results in a complex

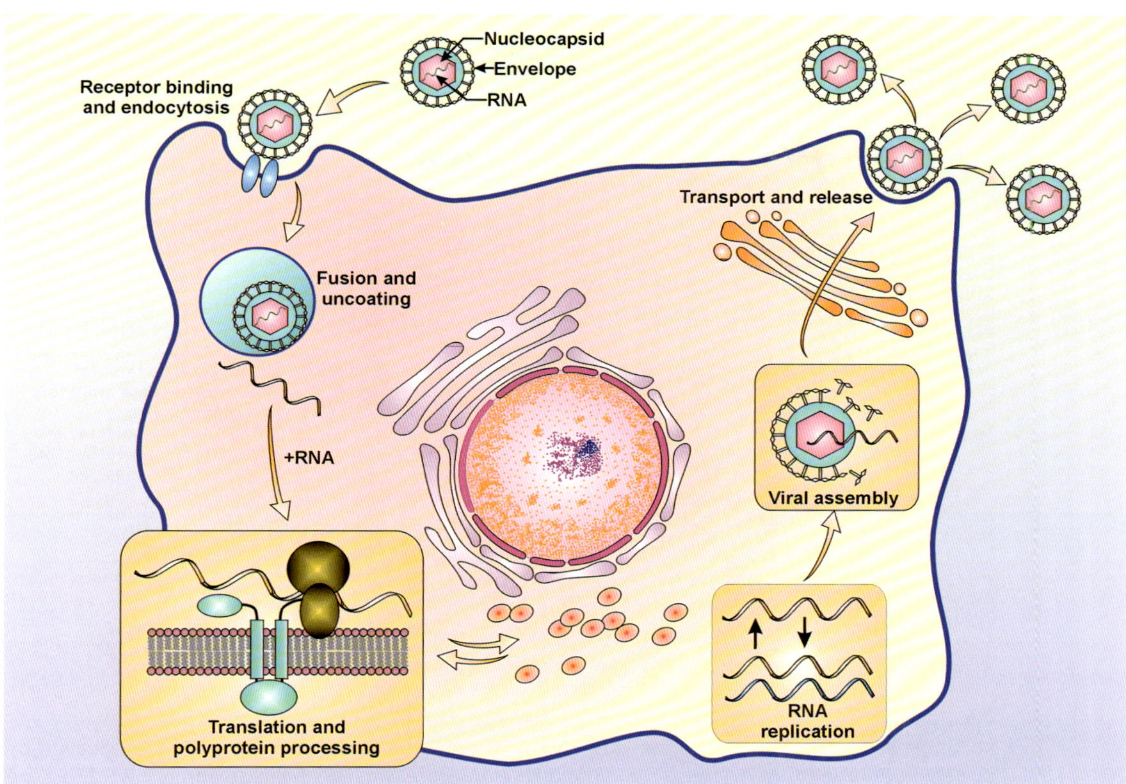

Fig. 1. HCV viral cycle. The HCV lifecycle starts with virion attachment to its specific receptor (not clearly identified). The HCV RNA genome serves as a template for viral replication and as a viral mRNA for virus production. It is translated into a polyprotein which is cleaved by proteases. Virus assembly then occurs. Each step of viral cycle is a potential target for drug development.

quasispecies population within each infected individual and enables rapid adaptation to changing environments. Six HCV genotypes and a large number of subtypes have been identified [3]. The HCV virion is made of a single-stranded positive RNA genome, contained in a capsid that is itself enveloped by a lipid bilayer with 2 different glycoproteins anchored.

The HCV lifecycle starts with virion attachment to its specific receptor (fig. 1). Several candidate molecules have been suggested to play a role in the receptor complex, including tetraspanin CD81, the scavenger receptor BI, the C-type lectin molecules DC-SIGN and L-SIGN and the low-density lipoprotein receptor [4]. Recently, the tight junction component claudin (CLDN-1) has been identified as additional key factors for HCV infection [5]. The CD81 partner EWI-2wint inhibits HCV entry, suggesting that, in addition to the presence of specific entry factors in the hepatocytes, the lack of a specific inhibitor can contribute to the hepatotropism of HCV [6].

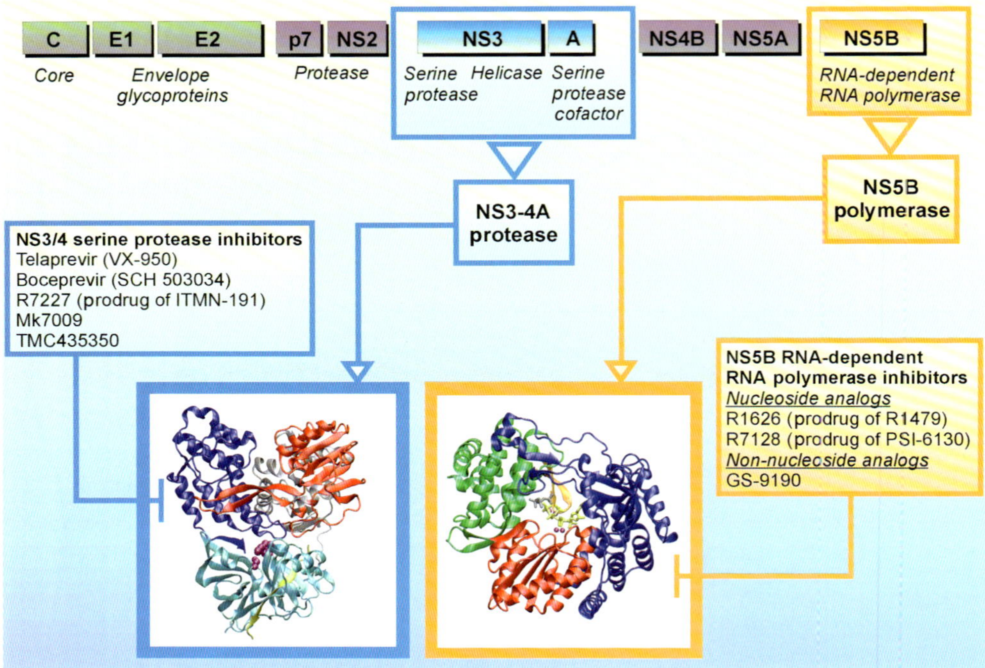

Fig. 2. HCV genome and potential drug discovery targets. HCV is an enveloped *Flavivirus* with a 9.6-kb single-strand RNA genome. The HCV RNA is translated into a polyprotein which is cleaved by proteases. All the HCV enzymes – NS2-3 and NS3-4A proteases, NS3 helicase and NS5B RdRp – are essential for HCV replication, and are therefore potential drug discovery targets. The knowledge of the structures of HCV protease and HCV polymerase has allowed inhibitors to these enzymes to be developed by structure-based drug design.

The HCV RNA genome serves as a template for viral replication and as a viral messenger RNA for viral production. It is translated into a polyprotein which is cleaved by proteases [7–9]. All the HCV enzymes, NS2-3 and NS3-4A proteases, NS3 helicase, and NS5B RdRp- are essential for HCV replication, and are therefore potential drug discovery targets (fig. 2). Since the structures of HCV protease and HCV polymerase were elucidated, numerous groups have used structure-based drug design to develop inhibitors of these enzymes.

Until 1999, the absence of a cell culture system supporting full replication of HCV and of convenient animal models limited the knowledge of the HCV life cycle and our ability to test antiviral molecules [8, 9]. The chimpanzee is the only currently available animal model for HCV infection [8]. The development of Huh 7, a subgenomic HCV RNA replicon capable of replication in the human hepatoma cell line, has been a significant advance [8, 9]. Recently, complete replication of HCV in cell culture has been achieved [10]. These models will improve our understanding of HCV replication and testing for antiviral molecules.

Interferon Signalling and HCV Infection

Endogenous type I IFNs are major antiviral cytokines. HCV infection may activate host signaling pathways that induce type I IFNs [11–13]. It should be remembered that the dsRNA virus induces host immune response; dsRNA is recognized as a pathogen-associated molecular pattern by pattern recognition receptors such as endosomal Toll-like receptor 3 (TLR-3) and cytosolic retinoic acid-inducible gene-I (RIG-I; fig. 3). Although HCV virus is a single-strand RNA, the fact that replication of the HCV genome is catalyzed by its RNA-dependent RNA polymerase NS5B, suggests that double-stranded RNA may be formed during the HCV life cycle and activate pattern recognition receptors. Interestingly, a recent study shows that specific homopolymeric RNA motifs within the genome of HCV and other RNA viruses can also act as a pathogen-associated molecular pattern substrate of RIG-I [11]. Activation of the TLR-3 pathway via the adaptor TRIF leads to phosphorylation of IFN regulatory factor-3 (IRF-3) and activation of transcription factors AP-1 and NF-κB. Phosphorylated IRF-3 forms a dimer, translocates into the nucleus, binds to DNA, and regulates the expression of IFN-β expression (IFN-β) in collaboration with AP-1 and NF-κB. After recognition of viral RNA, RIG-I and Mda5 recruit IFN-β promoter stimulator-1 (IPS-1, also known as Cardif) [12]. IPS-1 is localized into mitochondria and plays a critical role in the activation of IRF-7, IRF-3, and NF-κB. IRF-7 forms a dimer, translocates into the nucleus to induce IFN-α/β; homodimers of IRF-3 collaborate with NF-κB to induce IFN-β. IFN-α/β of autocrine/paracrine sources bind to a common receptor expressed at the cell surface. Receptor engagement causes the activation of Jak-STAT signalling which, together with IFN-stimulated growth factor (ISGF)3G/IRF-9, binds to IFN-stimulated response elements, thereby activating the transcription of IFN-α/β-inducible genes [13]. This results in the production of effector molecules, such as RnaseL, and protein kinase R, that will degrade viral RNAs and block their translation.

HCV RNA encodes specific proteins that may inhibit the induction of type I IFNs. An example is the NS3-4A protease of HCV, which blocks dsRNA-induced IFN production by interfering with IRF-3 phosphorylation [14]. NS3-4A mediates the cleaving of the C-terminal region of IPS-1/Cardif, causing disruption of NF-κB and IRF-3 activation, probably due to mislocalization of cleaved IPS-1/Cardif from mitochondria. NS3-4A also mediates TRIF proteolysis, suggesting multiple functions for this protease. Thus, HCV proteins may block both TLR- and RIG-I-Mda5-dependent signalling pathways to antagonize type I IFN induction. Thus, the NS3-4A protease is a dual therapeutic target, whose inhibition may block viral replication and restore IRF3-dependent control of HCV infection.

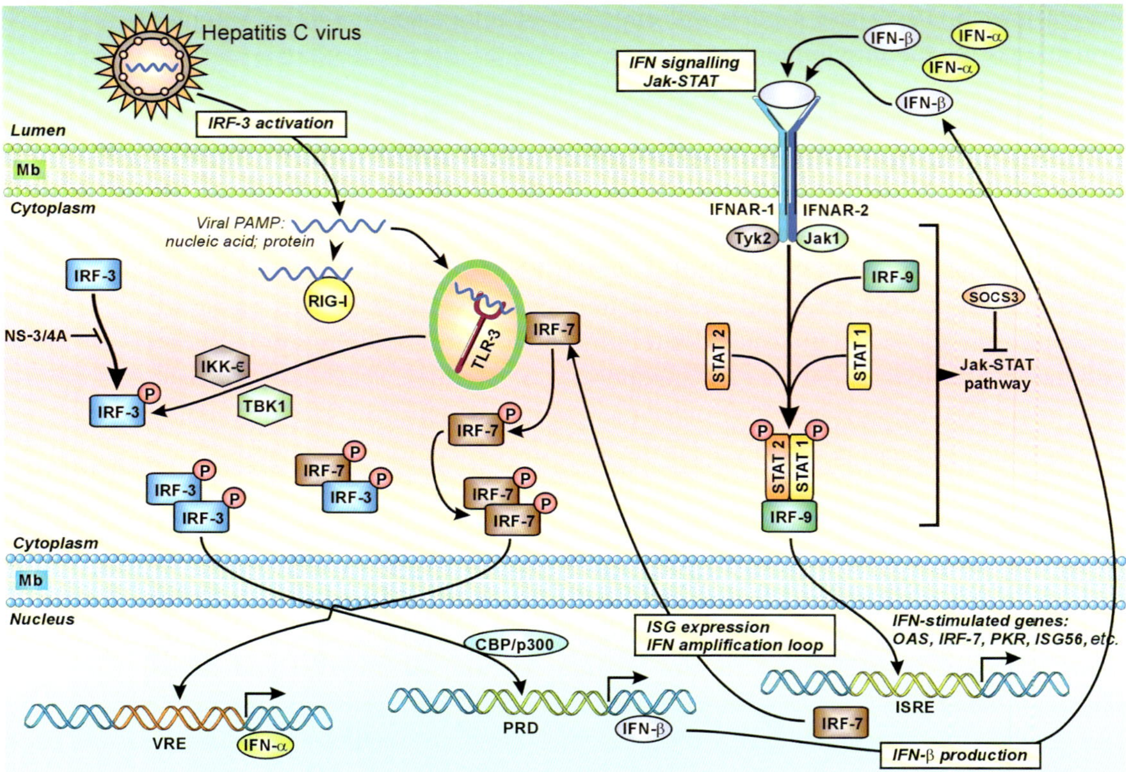

Fig. 3. HCV and immune response. Activation of TLR-3 leads to the recruitment of IκB kinase (IKK)-related kinases, TANK-binding kinase 1 (TBK1) and IKKε. These kinases, together with adaptators TANK and NAP1, catalyse the phosphorylation of IFN stimulatory factor-3 (IRF-3). Phosphorylated IRF-3 forms a dimer, translocates into the nucleus, binds to DNA in collaboration with transcription factor AP-1 and NF-κB and regulates the expression of IFN-β. The HCV NS34A serine protease may block the phosphorylation and effector action of IRF-3. After recognition of viral RNA, retinoic acid-inducible gene-I (RIG-I) and Mda5 recruit IFN-β promoter stimulator-1 (IPS-1) via CARD-CARD (caspase recruitment domain) interaction. IPS-1 is localized in the mitochondria and acts as an adaptator that plays a critical role in the activation of IRF-3 and IRF-7. IPS-1 is targeted and inactivated by the serine protease NS3-4A from HCV. IRF-7 forms a dimer and translocates into the nucleus to induce IFN-α/β. Endogenous IFN-α/β binds to a common receptor (IFNAR-1/2) expressed at the cell surface of target cells. Receptor engagement leads to recruitment of Tyk2 and Jak1. Together with IRF-9, the 2 kinases induce activation of STAT1 and STAT2 which, together with IFN-stimulated growth factor 3G (ISGF3G)/IRF-9, bind to *cis*-acting IFN-stimulated elements (ISREs), thereby activating the transcription of IFN-α/β-inducible genes such as protein kinase R, IL-8, OAS, etc. The HCV core protein has been shown to induce the expression of SOCS3 (suppressor of cytokine signalling 3), which can suppress Jak/STAT. Mb = Membrane; PAMP = Pathogen-associated molecular pattern; PRD = Positive regulatory domain; VRE = viral response element.

HCV-related effects may also attenuate IFN signalling. Proteins called suppressor of cytokine signalling (SOCS) are known to inhibit cytokine signalling via the Jak-STAT pathway. The HCV core protein has been shown to induce the expression of SOCS-3, which can suppress Jak-STAT signalling events and block the IFN-induced formation of ISGF3 [15]. HCV protein expression in liver cells is associated with induction of the protein inhibitor of activated STAT expression and inhibition of STAT function. Patients with chronic HCV infection have been shown to exhibit high levels of IL-8 in the liver [16]. The biological activity of IL-8 interferes with IFN signalling events that result in ISGF3 recruitment and function.

Together, these findings suggest that HCV modulation of IFN induction and signalling limits the expression of IFN-stimulated genes, allowing HCV to evade the antiviral actions of the host response.

Innate and Adaptive Immunity to HCV Infection

HCV is among the very few viruses that can lead to a chronic replicative infection in immunocompetent hosts. This implies the ability to escape the host's innate and adaptive immune response [17]. However, in about 30% of cases, the virus is cleared within 6 months following the acute and often asymptomatic infection [18], indicating its ability to induce an efficient anti-HCV immune response. It has also been shown that HCV-specific T cells can be detected in previously and chronically infected patients [18]. Hence, we propose that HCV infection is a fine balance between the pathogenicity and escape mechanisms of the virus, and the host's immune response. In this view, immune evasion and efficient anti-viral immunity represent 2 extreme scenarios leading to chronic infection or rapid eradication of the virus, respectively. We will describe some of the immune and virus-related factors that play a critical role in this balance and discuss potential explanations for the very diverse outcome of HCV infection in different individuals.

Innate immunity represents the first barrier against invading pathogens and has 2 main functions: (1) the early control of the infection through non-specific mechanisms; (2) the induction of an efficient Ag-specific adaptive immune response. Type I IFN response is an essential innate mechanism of host defense against viral infections. In the previous chapter, we described the molecular basis of IFN signaling and its potential modulation by viral proteins. When considering a global immune response, it is essential that IFN production occurs in the right place and at the right time, suggesting a tight spatio-temporal regulation. In HCV infection, the cellular source of IFN is not clearly defined. Infected hepatocytes can produce IFN, but in very low amounts. Plasmacytoid pre-dendritic cells (pDCs) are the professional type I IFN producing cells [19]. They can rapidly (within 6–12 h) produce large amounts of type I IFN in response to various viruses, such as herpes simplex or influenza [19]. It has been shown that the IFN response of pDCs to HCV was very poor as compared to herpes simplex virus and influenza [20], and possibly diminished by the HCV NS 5 protein [21]. However, it remains controversial whether the function of pDC from HCV-infected patients is impaired [22–24] or not [25]. Also, it is not clear whether pDCs are recruited into the liver in order to produce type I IFN at the primary site of the infection. NK cells represent another innate immune cell type that plays a critical role by killing virally-infected cells [26]. It was shown that ligation of CD81 by HCV-E2 protein inhibits NK cell function, potentially contributing to the persistence of the virus [27, 28]. NK cell inhibitory receptors also play a role in resolving HCV infection [29].

Dendritic cells (DCs) link innate and adaptive immunity by taking up microbial Ag and presenting them to naïve CD4 T cells in order to initiate Ag-specific immune responses [30]. It has been shown that the phagocytosis of HCV-infected apoptotic cells, but not HCV directly, can induce the maturation of human monocyte-derived DCs [31]. Studies in patients showed that DCs had impaired allostimmulatory function in HCV infection [32, 33], with increased IL-10 and decreased IL-12 production [34]. These results have been recently challenged by studies finding a normal phenotype and allostimulatory function in 13 HCV-infected patients [35], and in HCV-infected chimpanzees [36], supporting the view that immune deficiency in HCV infection may be specific of viral epitopes. HCV quasispecies were found in situ in liver DC [37] but their role in modulating DC function remains unclear. Because of their importance of initiating specific T cell responses, DC have been used as vaccines in different settings, such as tumours [38] and HIV infection [39]. DC loaded with HCV-specific Ag are potential candidates for HCV vaccination aiming at inducing a potent T cell immunity. Pre-clinical studies have yielded interesting results [40] but human trials are still lacking [41].

Cytotoxic T cell responses are essential in chronic viral infections. In chimpanzees, it was shown that an efficient cytotoxic T cell response was essential to eradicate the virus, whereas Ab titers were less predictive of the resolution of the infection [42]. Memory CD8 T cells also contribute to long-term protection against HCV [43]. Viral epitopes were presented to CD8 T cells by MHC class I molecules [44] and the presence of polyfunctional T cell responses was associated with effective control of viral replication [45]. Interestingly, CD4 T cell help was shown to be required for an efficient CD8 response [46].

Overall, immunity to HCV infection remains poorly characterized and a number of studies have been conducted in small number of patients or in in vitro systems of limited relevance, often yielding conflicting results. A potent cyto-toxic CD8 T cell response is certainly essential, but the innate parameters that control and initiate such responses in HCV-infected patients are not well characterized. Studies in large cohorts of patients in relevant models of acute or early-stage infection may be critical to improve our knowledge of HCV-specific immune responses and help design rational vaccination strategies.

Progression of Fibrosis in Hepatitis C

Liver fibrosis is the accumulation in excess of extracellular matrix proteins including collagen [47]. Fibrogenesis is a complex and dynamic process, mediated by necrosis, inflammation and the activation of stellate cells. The progression of fibrosis in chronic hepatitis C determines the ultimate prognosis and thus the need and urgency of therapy [48]. Liver biopsy remains the gold standard to assess fibrosis. Scoring systems (Knodell, Metavir, Ishak, etc.) provide a semi-quantitative assessment for individual clinical prognosis, studies and treatment trials [49]. The major factors associated with the progression of fibrosis are male gender, older age at infection, excess alcohol consumption and immuno-suppression [50, 51]. Insulin resistance may also play a role in the more rapid progression of fibrosis [52]. The natural history of liver fibrosis in chronic hepatitis C is influenced by both genetic and environmental factors.

In a previous study to identify molecular markers of prognosis in chronic hepatitis C, mRNA expression was quantified by real-time quantitative RT-PCR for a large number of selected genes in F2 (moderate fibrosis) specimens and compared to F1 (mild fibrosis) specimens [16]. Genes involved in the physiology of fibrosis were selected. Twenty-two genes were identified that were up-regulated in the F2 samples compared to F1 samples. These up-regulated genes were mainly those involved in encoding components of the cytoskeleton *(KRT19* and *SCG10)*, or chemokines/cytokines *(CXCL6, IL8, IL1A, IL2* and *CXCL10)*, or chemokine receptors *(CCR2, CXCR3* and *CXCR4)*, in extracellular matrix production *(COL1A1, CHI3L* and *SPP1)*, in extracellular matrix remodeling *(TIMP1, MMP7* and *MMP9)*, and in cell junctions *(ITGA2* and *CLDN4)*.

Interestingly, a recent study has demonstrated that histologically normal liver tissue obtained in 2 different ways (percutaneous or surgical liver biopsies) had different gene expression patterns although all specimens were histologically normal [53]. The most notable changes in gene expression occurred in the inflammatory response gene family. These results emphasize the importance of an adequate selection of histologically normal controls to prevent discordant or false results in gene expression profile analysis. Although it is difficult to state which is the best 'histologically normal' control, when performing studies, histological normal controls should always be obtained by the same technique as pathological samples. Finally, the careful selection of controls is crucial because inappropriate samples could lead to misinterpretation of results.

Identifying molecular markers of the progression of fibrosis has several clinical implications. First, many of the genes found up-regulated between mild and moderate fibrosis encode molecules secreted in the serum (cytokines). Therefore, looking for genes dysregulated in the liver can constitute a logical functional approach for the discovery of serum markers of the progression of fibrosis. Second, since a primary goal in the treatment of HCV infection is eradication of the virus and another is to stop the progression of fibrosis, molecular markers of the progression of fibrosis could help define new end-points during antiviral therapy. Therefore, gene changes could be new markers of the progression of fibrosis during antiviral treatment. Third, many of the up-regulated genes identified in this gene-expression study are potential molecular targets for the development of antifibrotic drugs.

Predictive Factors of Response to Treatment

A sustained virological response (SVR) rate of about 55% is obtained with the combination of pegylated IFNs (PEG-IFN) and ribavirin [54–57]. Long-term follow-up studies have shown that SVR is generally associated with clinical and histological improvement, eradication of HCV infection and progressive decrease of anti-HCV antibodies in most patients [58].

A significant number of patients will fail to respond to treatment, have virological relapse or experience side effects so significant that treatment must be stopped [59]. For reasons of both patient welfare and cost-effectiveness, it is therefore important to identify non-responding patients as early as possible (ideally at baseline before treatment). The probability of SVR essentially depends on genotype. Younger age, female gender, mild fibrosis and low viral load are also associated with a better response rate but to a lesser extent. In patients with HCV genotypes 2 or 3, the SVR rate is 80%; in genotype 1 about 50%.

Early virological response is the best predictive factor of a SVR [55]. A reduction in HCV RNA serum levels of less than 2 $\log_{10}$ copies/ml after the first 12 weeks of treatment as compared to baseline is clearly associated with almost no chance of SVR (negative predictive value 97–100%). A rapid virological response (undetectable HCV RNA at week 4) seems to be the best predictor of treatment outcome in patients with chronic hepatitis C [60].

In patients infected with HCV genotype 1, the rates of sustained virologic response and tolerability did not differ significantly between the 2 available PEG-IFN-ribavirin regimens or between the 2 doses of PEG-IFNα-2b [61]. Among 3,070 patients, rates of sustained virologic response were similar among the regimens: 39.8% with standard-dose PEG-IFNα-2b; 38.0% with low-dose PEG-IFNα-2b, and 40.9% with PEG-IFNα-2a. Interestingly, among the 10% of patients with undetectable HCV RNA levels at treatment week 4, 86% had a sustained virologic response. The 24% of patients with a minimal decline in HCV RNA level (decline of $<1 \log_{10}$ IU from the baseline value) at treatment week 4 had a probability of sustained virologic response of less than 5%. Thus, virologic response at treatment week 4 is an important predictor of sustained virologic response. HCV RNA levels should be routinely assessed at this time point.

It was recently reported that non-responders might be re-treated with higher dose or longer duration of PEG-IFN-α 2b plus ribavirin. While the overall rate of SVR is low, and increasing the dose has no benefit, it seems that treating for 72 weeks compare to 48 weeks increases the SVR rate in a particular group of patients [62].

Genome-Wide Association Studies Identified SNPs in the IL28B Region Associated with Response to Treatment

Three independent genome-wide association studies reported SNP in the IL28B region associated with response to treatment [63–65]. All these patients were infected by genotype 1, and received the standard of care (PEG-IFN + ribavirine). In these 3 studies, the HCV-infected populations were from different countries and origins (European, African American, Australian, Japanese).

Ge et al. [63] analysed 1,137 patients with HCV genotype 1 infection, and they identified several SNPs near the IL28B gene on chromosome 19 that were significantly more common in responders than in non-responders.

Interestingly, Suppiah et al. [64] and Tanaka et al. [65] identified RS 8099917 (located ~8 kb upstream of IL28B) as the variant most strongly associated with SVR. A study by Thomas et al. [66] reports that the same variant described by Ge et al. is also associated with spontaneous clearance of HCV. IFN-λs, including IFN-λ1, IFN-λ2 and IFN-λ3, also known as IL-29, IL-28A or IL-28B, are a newly described group of cytokines distantly related to the type I IFNs and IL-10 family members [67]. The IFN-λR complex consists of a unique ligand-binding chain, IFN-λR1 (also designated IL-28Rλ), and an accessory chain, IL-10R2, which is shared with receptors for IL-10-related cytokines. IFN-λs signal through the IFN-λR and activate the JAK-STAT and MAPK pathways to induce antiviral, antiproliferative, anti-tumour and immune responses.

Although all of the identified variants in the 3 studies lie in or near the IL28B gene, none of them has an obvious effect on the function of this gene, which encodes IFN-λ3. The IFN-λ proteins seem to have lower antiviral activity than IFN-α in vitro [68]. IFN-λ1 has been shown to exhibit dose- and time-dependent HCV inhibition, induce increases in levels of IFN-stimulated genes, and enhance the antiviral efficacy of IFN-α [69].

Molecular Signatures of Response in Liver

Liver gene expression profiling has recently been applied to chronic hepatitis C to determine response to therapy. Knowledge of the antiviral mechanisms of IFN is crucial for the discovery of new treatments (fig. 3).

In a recent study, a selection of genes associated with liver gene expression dysregulation during HCV infection was studied by large scale RT-PCR assay according to response to treatment [70]. Supervised class prediction analysis identified a 2-gene *(IFI27* and *CXCL9)* signature, which accurately predicted treatment response in 79.3% of patients from the validation set, with a predictive accuracy of 100% and of 70% in non-responders and SVRs, respectively. The expression profiles of responder-relapsers did not differ significantly from those of non-responders and SVRs, and 73% of them were predicted as SVRs with the 2-gene classifier.

Table 1. List of the genes that differ between non-responders (NR) and sustained virological responders (SVR) (training set)

Gene symbol	Family	Protein encoded	Non-response:SVR ratio[a]
IFI-6-16	IFN-inducible protein	IFN-α-inducible protein 3	3.5
IFI27	IFN-inducible protein	IFN-α-inducible protein 27	4.2
ISG15	IFN-inducible protein	IFN-α-inducible protein 2	3.7
MX1	IFN-inducible protein	activating transcription factor 6	2.7
HERC5	IFN-inducible protein	Hect domain and RLD 5	2.2
TGFB2	growth factor	TGF-β2	2.7
OAS2	IFN-inducible protein	29-59-oligoadenylate synthetase 2	1.8
VEGFD	angiogenesis	Vascular endothelial growth factor D	2.4
IL8	interleukin	Interleukin 8	3.2
IFIT1	IFN-inducible protein	IFN-induced protein with tetratricopeptide repeats 1	55.3

[a] Gene expression ratios were compared among non-responder and SVR liver gene expression values.

In conclusion, non-responders and sustained virological responders have different gene expression profiles prior to treatment. The most notable changes in gene expression were mainly observed in the IFN-stimulated genes (table 1). We were able to predict treatment response with a 2-genes signature *(IFI27* and *CXCL9)* in 2 independent groups of patients (training and validation set). Most relapsers clustered with SVR patients. Interestingly, the baseline liver levels of expression of IFN-stimulated genes were higher in non-responders than in SVRs. The failure to respond to exogenous PEG-IFN could indicate a blunted response to IFN. This suggests that IFN-stimulated genes are already maximally induced in patients who failed to respond. Genes included in the signature encode molecules secreted in the serum and provide a logical functional approach for the development of serum markers to predict response to treatment.

In a recent study, the authors have identified the gene expression patterns in peripheral blood mononuclear cells during IFN therapy and confirmed the up-regulation of genes thought to be IFN-stimulated genes or involved in antigen processing and presentation [71].

Interestingly, data in HCV-infected chimpanzees indicate a predominantly defective hepatic response to IFN, which is probably mediated through the activation of SOCS3 and may explain the lack of response to IFN-based therapy in many HCV patients [72].

Many of the genes that are up-regulated in non-responders compared to responders encode molecules secreted in the serum (cytokines/chemokines). Thus, these targets represent a logical and functional approach for the development of serum markers as predictors of response to treatment. In a recent study, we examined the levels of chemokines that bind to CXC chemokine receptor 3 (CXCR3) in order to determine whether these chemokines play a role in the failure of the immune system to clear HCV infection [73]. Levels of CXCL10 and CXCL9 decreased during successful antiviral therapy while CXCL11 did not decrease significantly within the first 6 months after therapy. Baseline levels of CXCL10 were highest in HCV patients who did not respond to therapy. These results suggest that plasma concentrations of CXCL10 may be a predictor of response to PEG-IFN ± ribavirin. Interestingly, we identified CXCL9 in our 2-gene signature, and that CXCL9, CXCL10 and CXCL11 have the same specific receptor CXCR3 [70].

Standard of Care: Combined PEG-IFN Plus Ribavirin

Management of patients with chronic hepatitis C should be global. All factors associated with a rapid progression of fibrosis should be identified, together with those affecting the response to treatment (such as excess alcohol consumption, obesity, insulin resistance), and ultimately be controlled.

The main treatment goal in chronic hepatitis C is the prevention of cirrhosis and hepatocellular carcinoma by eradicating the virus. Recently, advances have been made in treatment with the combination of PEG-IFN and ribavirin. At present, in a patient with hepatitis C, therapy results in a sustained response in approximately 55% of cases.

In patients with HCV genotypes 2 or 3, the SVR rate is about 80% and in genotype 1 approximately 50%. Based on existing results, the sustained virological response with this treatment option appears to be long lasting, to be associated with a histological benefit and is also probably associated with a reduction in the risk of cirrhosis and hepatocellular carcinoma.

Specifically Targeted Antiviral Therapy for HCV

Protease Inhibitors

The development of new molecules such as viral enzyme inhibitors (proteases and polymerases) is necessary. Understanding the tridimensional structures of viral proteases and helicases (fig. 2) was an important step in developing new drugs because specific inhibitors could be developed for these enzymes. In 2003–2004, a first protease inhibitor that specifically blocks viral C replication in the replicon model was shown to be effective in human [74, 75]. Unfortunately, the development of this molecule was discontinued because of cardiac toxicity observed in the chimpanzee.

Table 2. Protease and polymerase inhibitors in development for the treatment of hepatitis C

Drug name	Drug class	Preclinical	Phase I	Phase II	Phase III
Telaprevir (VX-950) (Janssen, Tibotec)	NS3/4 serine protease inhibitors				
Boceprevir (SCH 503034) (Schering-Plough)	NS3/4 serine protease inhibitors				
BI201335 (Bohringer)	NS3/4 serine protease inhibitors				
TMC435350 (Tibotec, Medivir)	NS3/4 serine protease inhibitors				
R7227 (prodrug of ITMN-191) (Roche, Intermune)	NS3/4 serine protease inhibitors				
MK7009 (Merck)	NS3/4 serine protease inhibitors				
R1626 (prodrug of R1479) (Roche)	NS5B RNA-dependent RNA-polymerase inhibitors (nucleoside analogues)				
R7128 (prodrug of PSI-6130) (Roche)	NS5B RNA-dependent RNA-polymerase inhibitors (nucleoside analogues)				
GS-9190 (Gilead Sciences)	NS5B RNA-dependent RNA-polymerase inhibitors (non-nucleoside inhibitors)				
A-831 (Astra Zeneca, Arrow)	NS5A inhibitors				

Currently, several specific targeted agents against HCV are in development (table 2). Telaprevir and boceprevir are the 2 more advanced protease inhibitors.

Telaprevir. The antiprotease NS3-NS4A telaprevir is being developed by Vertex and Tibotec. In this study, the anti-protease has been combined to PEG-IFN ± ribavirin. Adding telaprevir to the standard of care presents 2 significant advantages: (1) it increases SVR rates from 50% to 70% in patient with genotype I, and (2) the duration of the treatment can be reduced from 48 to 24 weeks. Responses rates were lowest with the regimen that did not include ribavirin. However, the telaprevir regimen is related to a higher rate of discontinuation because of adverse events (mainly cutaneous rash) [76, 77].

Boceprevir. Developed by Schering-Plough, boceprevir is a small molecule, which is a specific inhibitor of the viral protease NS3-NS4A. The association of PEG-IFN 2b with boceprevir resulted in a marked viral load reduction of approximately 2.5 log after 13 days of treatment in non-responders to PEG-IFN ± Ribavirin [78, 79].

When boceprevir is combined with standard of care, it appears safe and substantially improves SVR rates. In this trial, 48 weeks of boceprevir treatment almost doubled the SVR rate compared to the current standard of care. A 4-week lead-in with standard of care prior to the addition of boceprevir appeared to reduce the incidence of viral breakthrough. The most common adverse events reported in the boceprevir arms were fatigue, anaemia, nausea and headache.

Polymerase Inhibitors

Polymerase inhibitors interfere with viral replication by binding to the NS5B RNA-dependent RNA polymerase. Two types of polymerase inhibitors have been developed: (1) nucleoside analogues and (2) non-nucleosides inhibitors. Nucleoside analogues act as chain terminators: they interfere with initiation of RNA transcription and elongation. Nucleoside analogues target the active site of HCV polymerase. In contrast, non-nucleoside inhibitors have been designed to bind to several discrete sites on HCV polymerase. The resistance profiles of nucleoside analogues, non-nucleoside inhibitors and protease inhibitors are all distinct from each other. Thus, it is possible that agents from different classes will act in a complementary fashion to increase their efficiency and thus prevent the development of resistance.

Conclusion

In chronic hepatitis C, despite progress performed with the standard of care (PEG-IFN + ribavirin), treatment failure still occurs in about half of the patients. Furthermore, therapy results in several side effects and high cost. The development of new drugs is mandatory and will depend on a better knowledge of the viral cycle and mechanisms of non response.

Promising results have been reported with 2 protease inhibitors (telaprevir and boceprevir) which are currently in phase III trial. Several other protease and polymerase inhibitors are under development.

In the near future, it is likely that IFN-based therapy plus ribavirin will remain the backbone of the treatment of chronic hepatitis C. PEG-IFN and ribavirin are needed in order to prevent HCV resistance to STAT-C drugs, and subsequently increase SVR. Genotypic and phenotypic resistance tests will also enter the therapeutic arena. Once several STAT-C agents become available, new treatment strategies will include the combination of several drugs with different mechanisms of action (protease inhibitors plus polymerase inhibitors) that could hopefully spare the use of PEG-IFN ± ribavirin. Future treatments might be combinations of antiviral drugs that have additive or synergistic potency, do not induce cross resistance, and have a good safety profile.

Acknowledgment

We thank Jean-Pierre Laigneau (from INSERM, Centre de Recherche Bichat-Beaujon CRB3, Paris, France) for the medical illustrations.

References

1 Seeff LB: Natural history of chronic hepatitis C. Hepatology 2002;36:S35–S46.
2 Choo QL, Kuo G, Weiner AJ, et al: Isolation of a cDNA clone derived from a blood-borne non-A, non-B viral hepatitis genome. Science 1989;244:359–362.
3 Simmonds P, Bukh J, Combet C, et al: Consensus proposals for a unified system of nomenclature of hepatitis C virus genotypes. Hepatology 2005;42:962–973.
4 Dubuisson J, Helle F, Cocquerel L: Early steps of the hepatitis C virus life cycle. Cell Microbiol 2008;10:821–827.
5 Evans MJ, von Hahn T, Tscherne DM, et al: Claudin-1 is a hepatitis C virus co-receptor required for a late step in entry. Nature 2007;446:801–805.
6 Rocha-Perugini V, Montpellier C, Delgrange D, et al: The CD81 partner EWI-2wint inhibits hepatitis C virus entry. PLoS ONE 2008;3:e1866.
7 Lanford RE, Bigger C, Bassett S, et al: The chimpanzee model of hepatitis C virus infections. ILAR J 2001;42:117–126.
8 Lohmann V, Korner F, Koch J, et al: Replication of subgenomic hepatitis C virus RNAs in a hepatoma cell line. Science 1999;285:110–113.
9 Blight KJ, Kolykhalov AA, Rice CM: Efficient initiation of HCV RNA replication in cell culture. Science 2000;290(5498):1972–1974.
10 Wakita T, Pietschmann T, Kato T,et al: Production of infectious hepatitis C virus in tissue culture from a cloned viral genome. Nat Med 2005;11:791–796. Erratum in: Nat Med 2005;11:905.
11 Saito T, Owen DM, Jiang F, et al: Innate immunity induced by composition-dependent RIG-I recognition of hepatitis C virus RNA. Science 2008;24;454:523–527.
12 Saito T, Hirai R, Loo YM, et al: Regulation of innate antiviral defenses through a shared repressor domain in RIG-I and LGP2. Proc Natl Acad Sci USA 2007;104:582–587.
13 Gale M Jr, Foy EM: Evasion of intracellular host defence by hepatitis C virus. Nature 2005;436:939–945. Erratum in: Nature 2005;437:290.
14 Foy E, Li K, Wang C, Sumpter R Jr, et al: Regulation of interferon regulatory factor-3 by the hepatitis C virus serine protease. Science 2003;300:1145–1148.
15 Bode JG, Ludwig S, Ehrhardt C, et al: IFN-alpha antagonistic activity of HCV core protein involves induction of suppressor of cytokine signaling-3. FASEB J 2003;17:488–490.
16 Asselah T, Bièche I, Laurendeau I, et al: Liver gene expression signature of mild fibrosis in patients with chronic hepatitis C. Gastroenterology 2005;129:2064–2075.
17 Bode JG, Brenndörfer ED, Häussinger D: Hepatitis C virus (HCV) employs multiple strategies to subvert the host innate antiviral response. Biol Chem 2008;389:1283–1298.
18 Bowen DG, Walker CM: Adaptive immune responses in acute and chronic hepatitis C virus infection. Nature 2005;436:946–952.
19 Liu YJ: IPC: professional type 1 interferon-producing cells and plasmacytoid dendritic cell precursors. Annu Rev Immunol 2005;23:275–306.
20 Gondois-Rey F, Dental C, Halfon P, et al: Hepatitis C virus is a weak inducer of interferon alpha in plasmacytoid dendritic cells in comparison with influenza and human herpesvirus type-1. PLoS ONE 2009;4:e4319.
21 Amjad M, Abdel-Haq N, Faisal M, et al: Decreased interferon-alpha production and impaired regulatory function of plasmacytoid dendritic cells induced by the hepatitis C virus NS 5 protein. Microbiol Immunol 2008;52:499–507.
22 Goutagny N, Vieux C, Decullier E, et al: Quantification and functional analysis of plasmacytoid dendritic cells in patients with chronic hepatitis C virus infection. J Infect Dis 2004;189:1646–1655.
23 Szabo G, Dolganiuc A: Subversion of plasmacytoid and myeloid dendritic cell functions in chronic HCV infection. Immunobiology 2005;210:237–247.
24 Dolganiuc A, Chang S, Kodys K, et al: Hepatitis C virus (HCV) core protein-induced, monocyte-mediated mechanisms of reduced IFN-alpha and plasmacytoid dendritic cell loss in chronic HCV infection. J Immunol 2006;177:6758–6768.
25 Decalf J, Fernandes S, Longman R, et al: Plasmacytoid dendritic cells initiate a complex chemokine and cytokine network and are a viable drug target in chronic HCV patients. J Exp Med 2007;204:2423–2437.
26 Vivier E, Tomasello E, Baratin M, et al: Functions of natural killer cells. Nat Immunol 2008;9:503–510.
27 Tseng CT, Klimpel GR: Binding of the hepatitis C virus envelope protein E2 to CD81 inhibits natural killer cell functions. J Exp Med 2002;195:43–49.
28 Crotta S, Stilla A, Wack A, et al: Inhibition of natural killer cells through engagement of CD81 by the major hepatitis C virus envelope protein. J Exp Med 2002;195:35–41.
29 Khakoo SI, Thio CL, Martin MP, et al: HLA and NK cell inhibitory receptor genes in resolving hepatitis C virus infection. Science 2004;305:872–874.
30 Steinman RM, Banchereau J: Taking dendritic cells into medicine. Nature 2007;449:419–426.

31 Ebihara T, Shingai M, Matsumoto M, et al: Hepatitis C virus-infected hepatocytes extrinsically modulate dendritic cell maturation to activate T cells and natural killer cells. Hepatology 2008;48:48–58.
32 Bain C, Fatmi A, Zoulim F, et al: Impaired allostimulatory function of dendritic cells in chronic hepatitis C infection. Gastroenterology 2001;120: 512–524.
33 Sarobe P, Lasarte JJ, Casares N, et al: Abnormal priming of CD4(+) T cells by dendritic cells expressing hepatitis C virus core and E1 proteins. J Virol 2002;76:5062–5070.
34 Dolganiuc A, Kodys K, Kopasz A, et al: Hepatitis C virus core and nonstructural protein 3 proteins induce pro- and anti-inflammatory cytokines and inhibit dendritic cell differentiation. J Immunol 2003;170:5615–5624.
35 Longman RS, Talal AH, Jacobson IM, et al: Presence of functional dendritic cells in patients chronically infected with hepatitis C virus. Blood 2004;103:1026–1029.
36 Larsson M, Babcock E, Grakoui A, et al: Lack of phenotypic and functional impairment in dendritic cells from chimpanzees chronically infected with hepatitis C virus. J Virol 2004;78:6151–6161.
37 Laporte J, Bain C, Maurel P: Differential distribution and internal translation efficiency of hepatitis C virus quasispecies present in dendritic and liver cells. Blood 2003;101:52–57.
38 Palucka AK, Ueno H, Fay JW, et al: Taming cancer by inducing immunity via dendritic cells. Immunol Rev 2007;220:129–150.
39 Rinaldo CR: Dendritic cell-based human immunodeficiency virus vaccine. J Intern Med 2009;265:138–158.
40 Gehring S, Gregory SH, Wintermeyer P, et al: Generation of immune responses against hepatitis C virus by dendritic cells containing NS5 protein-coated microparticles. Clin Vaccine Immunol 2009; 16:163–171.
41 Irshad M, Khushboo I, Singh S, et al: Hepatitis C virus (HCV): a review of immunological aspects. Int Rev Immunol 2008;27:497–517.
42 Cooper S, Erickson AL, Adams EJ, et al: Analysis of a successful immune response against hepatitis C virus. Immunity 1999;10:439–449.
43 Shoukry NH, Grakoui A, Houghton M, et al: Memory CD8+ T cells are required for protection from persistent hepatitis C virus infection. J Exp Med 2003;197:1645–1655.
44 Kowalski H, Erickson AL, Cooper S, et al: Patr-A and B, the orthologues of HLA-A and B, present hepatitis C virus epitopes to CD8+ cytotoxic T cells from two chronically infected chimpanzees. J Exp Med 1996;183:1761–1775.
45 Ciuffreda D, Comte D, Cavassini M, et al: Polyfunctional HCV-specific T-cell responses are associated with effective control of HCV replication. Eur J Immunol 2008;38:2665–2677.
46 Grakoui A, Shoukry NH, Woollard DJ, et al: HCV persistence and immune evasion in the absence of memory T cell help. Science 2003;302:659–662.
47 Friedman SL: Hepatic stellate cells: protean, multifunctional, and enigmatic cells of the liver. Physiol Rev 2008;88:125–172.
48 Marcellin P, Asselah T, Boyer N: Fibrosis and disease progression in hepatitis C, Hepatology 2002; 36:S47–S56.
49 Bedossa P, Poynard T: An algorithm for the grading of activity in chronic hepatitis C. Hepatology 1996; 24:289–293.
50 Asselah T, Rubbia-Brandt L, Marcellin P, et al: Steatosis in chronic hepatitis C: why does it really matter? Gut 2006;55:123–130.
51 Peffault de Latour R, Levy V, Asselah T, et al: Long-term outcome of hepatitis C infection after bone marrow transplantation. Blood 2004;103:1618–1624.
52 Moucari R, Asselah T, Cazals-Hatem D, et al: Insulin resistance in chronic hepatitis C: association with genotypes 1 and 4, serum HCV RNA level, and liver fibrosis. Gastroenterology 2008;134:416–423.
53 Asselah T, Bièche I, Laurendeau I, et al: Significant gene expression differences in histologically 'normal' liver biopsies: implications for control tissue. Hepatology 2008;48:953–962.
54 Manns MP, McHutchison JG, Gordon SC, et al: Peginterferon alfa-2b plus ribavirin compared with IFN-2b plus ribavirin for initial treatment of chronic hepatitis C :a randomised trial. Lancet 2001;358:958–965.
55 Fried MW, Shiffman ML, Reddy KR, et al: Peginterferon alfa-2a plus ribavirin for chronic hepatitis C virus infection. N Engl J Med 2002;347:975–982.
56 Hadziyannis SJ, Sette H Jr, Morgan TR, et al: Peginterferon-alpha2a and ribavirin combination therapy in chronic hepatitis C: a randomized study of treatment duration and ribavirin dose. Ann Intern Med 2004;140:346–355.
57 Bronowicki JP, Ouzan D, Asselah T, et al: Effect of ribavirin in genotype 1 patients with hepatitis C responding to pegylated interferon alfa-2a plus ribavirin. Gastroenterology 2006;131:1040–1048.
58 Maylin S, Martinot-Peignoux M, Moucari R, et al: Eradication of hepatitis C virus in patients successfully treated for chronic hepatitis C. Gastroenterology 2008;135:821–829.

59 Marcellin P, Lau GK, Zeuzem S, et al: Comparing the safety, tolerability and quality of life in patients with chronic hepatitis B vs. chronic hepatitis C treated with peginterferon alpha-2a. Liver Int 2008; 28:477–485.
60 Maylin S, Martinot-Peignoux M, Ripault MP, et al: Sustained virological response is associated with clearance of hepatitis C virus RNA and a decrease in hepatitis C virus antibody. Liver Int 2009;29:511–517.
61 McHutchison JG, Lawitz EJ, Shiffman ML, et al: Peginterferon alfa-2b or alfa-2a with ribavirin for treatment of hepatitis C infection. N Engl J Med 2009;361:580–593.
62 Jensen DM, Marcellin P, Freilich B, et al: Re-treatment of patients with chronic hepatitis C who do not respond to peginterferon-alpha2b: a randomized trial. Ann Intern Med 2009;150:528–540.
63 Ge D, Fellay J, Thompson AJ, et al: Genetic variation in IL28B predicts hepatitis C treatment-induced viral clearance. Nature 2009;461:399–401.
64 Suppiah V, Moldovan M, Ahlenstiel G, et al: IL28B is associated with response to chronic hepatitis C interferon-alpha and ribavirin therapy. Nat Genet 2009. DOI: 10.1038/ng.447.
65 Tanaka Y, Nishida N, Sugiyama M, et al: Genome-wide association of IL28B with response to pegylated interferon-alpha and ribavirin therapy for chronic hepatitis C. Nat Genet 2009. DOI: 10.1038/ng.449.
66 Thomas DL, Thio CL, Martin MP, et al: Genetic variation in IL28B and spontaneous clearance of hepatitis C virus. Nature 2009. DOI: 10.1038/nature08463.
67 Li M, Liu X, Zhou Y, Su SB: Interferon-lambdas: the modulators of antivirus, antitumor, and immune responses. J Leukoc Biol 2009;86:23–32.
68 Sheppard P, Kindsvogel W, Xu W, et al: IL-28, IL-29 and their class II cytokine receptor IL-28R. Nature Immunol 2003;4:63–68.
69 Marcello T, Grakoui A, Barba-Spaeth G, et al: Interferons alpha and lambda inhibit hepatitis C virus replication with distinct signal transduction and gene regulation kinetics. Gastroenterology 2006;131:1887–1898.
70 Asselah T, Bieche I, Narguet S, et al: Liver gene expression signature to predict response to pegylated interferon plus ribavirin combination therapy in patients with chronic hepatitis C. Gut 2008;57: 516–524.
71 Ji X, Cheung R, Cooper S, et al: Interferon alfa regulated gene expression in patients initiating interferon treatment for chronic hepatitis C. Hepatology 2003;37:610–621.
72 Huang Y, Feld JJ, Sapp RK, et al: Defective hepatic response to interferon and activation of suppressor of cytokine signaling 3 in chronic hepatitis C. Gastroenterology 2007;132:733–744.
73 Butera D, Marukian S, Iwamaye AE, et al: Plasma chemokine levels correlate with the outcome of antiviral therapy in patients with hepatitis C. Blood 2005;106:1175–1182.
74 Lamarre D, Anderson PC, Bailey M, et al: An NS3 protease inhibitor with antiviral effects in humans infected with hepatitis C virus. Nature 2003;426:186–189.
75 Hinrichsen H, Benhamou Y, Wedemeyer H, et al: Short-term antiviral efficacy of BILN 2061, a hepatitis C virus serine protease inhibitor, in hepatitis C genotype 1 patients. Gastroenterology 2004;127: 1347–1355.
76 Hézode C, Forestier N, Dusheiko G, et al: Telaprevir and peginterferon with or without ribavirin for chronic HCV infection. N Engl J Med 2009;360: 1839–1850.
77 McHutchison JG, Everson GT, Gordon SC, et al: Telaprevir with peginterferon and ribavirin for chronic HCV genotype 1 infection. N Engl J Med 2009;360:1827–1838.
78 Sarrazin C, Rouzier R, Wagner F, et al: SCH 503034, a novel hepatitis C virus protease inhibitor, plus pegylated interferon alpha-2b for genotype 1 nonresponders. Gastroenterology 2007;132:1270–1278.
79 Kwo P, Lawitz EJ, McCone J, et al: HCV SPRINT-1:Boceprevir plus peginterferon alfa-2b/ribavirin for treatment of genotype 1 chronic hepatitis C in previously untreated patients. Hepatology 2008;48: LB16.

Dr. Tarik Asselah
Service d'Hépatologie, Hôpital Beaujon
100 Boulevard du Général Leclerc
FR–92110 Clichy (France)
Tel. +33 140 875579, Fax +33 147 309440, E-Mail tarik.asselah@bjn.aphp.fr

Mayerle J, Tilg H (eds): Clinical Update on Inflammatory Disorders of the Gastrointestinal Tract.
Front Gastrointest Res. Basel, Karger, 2010, vol 26, pp 59–71

Clinical Update on Inflammatory Disorders of the GI Tract: Liver Transplantation

Olivier de Rougemont · Philipp Dutkowski · Pierre-Alain Clavien

Department of Visceral and Transplantation Surgery, Swiss Hepato-Pancreatico-Biliary and Transplant Center, University Hospital Zurich, Zürich, Switzerland

Abstract

Inflammatory bowel disease (IBD) is well known to be associated with primary sclerosing cholangitis (PSC) and autoimmune hepatitis (AIH). In both of these types of autoimmune liver disease liver transplantation remains a well accepted treatment modality. Importantly, PSC patients are evaluated early for liver transplantation because of their high risk of developing cholangiocarcinoma. This strategy is based on new combined strategies with excellent 5-year survival in selected PSC patients with cholangiocarcinoma. However, recurrence or de novo development of PSC and AIH as well as repeated IBD after liver transplantation has long been a subject of debate. As most of these liver transplant recipients survive long term, the recurrence of disease is becoming an important cause of morbidity and mortality. In this review we systematically report on indications for liver transplantation in PCS and AIH. We also refer to current knowledge about clinical course, management, long term outcome and risk factors for disease recurrence.

Inflammatory bowel disease (IBD) appears as a chronic disease of the digestive tract and is thought to result from inappropriate and ongoing activation of the mucosal immune system. IBD is usually divided into 2 distinct entities: ulcerative colitis and Crohn's disease.

Primary sclerosing cholangitis (PSC) is a chronic cholestatic disease of unknown aetiology (although it is possibly autoimmune), that is characterized by inflammation and fibrosis of the intrahepatic and extrahepatic bile ducts. Autoimmune hepatitis (AIH) is a predominantly periportal hepatitis, usually with hypergammaglobulinaemia and tissue autoantibodies, and it generally responds to immunosuppressive therapy. Both PSC and AIH are associated with an increased prevalence of IBD (PSC 55–75%), with mild but extensive colitis being the most common presentation [1]. The prevalence of IBD in AIH is much lower [2].

While end-stage liver disease in patients with AIH or PSC is an accepted indication for liver transplantation, exacerbation of pre-existent and the development of de novo IBD have been described before and after transplantation because of – or despite – the use of potent immunosuppressive regimes. In this review we will give an overview about issues prior to transplantation in PSC and AIH. We will also discuss transplant surgery, post-transplantation complications, post-transplantation survival, and risk factors for recurrent disease.

Primary Sclerosing Cholangitis

Issues Prior to Transplantation

Until 30 years ago, PSC was considered a rare disease, but today it is one of the most common indications for liver transplantation in adults, a change that was – in part – brought about by the use of endoscopic retrograde cholangiopancreatography. Characteristically this chronic cholestatic liver disease leads to bilary duct inflammation resulting in fibrotic strictures and obliteration of extrahepatic and/or intrahepatic bile ducts. Biliary cirrhosis and portal hypertension follow and progress to end-stage liver disease. Cholangiographic findings are diagnostic for PSC, which is predominant in males [3]. Between 70 and 80% of cases are associated with IBD [4], and approximately 5% of patients with ulcerative colitis have PSC [5]. The disease can also appear de novo following liver transplantation [6]. At this time there are no reports of effective medical therapies for PSC. Medical approaches have principally focused on choleretic, immunosuppressive and antifiberogenic agents. A recent randomized controlled trial of high-dose ursodeoxycholic acid treatment showed an improvement in serum liver test but, as with other reports, did not improve survival [7]. Reconstructive biliary tract procedures and endoscopically placed stents may relieve symptoms associated with obstruction but a long-term effect on survival cannot be achieved here either. Ultimately, liver transplantation remains the only option for patients with PSC. A number of specific issues are well known prior to transplantation, which all should be taken into consideration for the decision to proceed to liver transplantation.

Bacterial Cholangitis

Impaired bile flow due to diffuse intrahepatic fibrotic stricturing of the biliary tree frequently causes episodes of bacterial cholangitis [8], a complication that occurs in about one third of patients. Broad-spectrum antibiotic coverage with alternating drugs for Gram-negative bacteria and *Enterococcus* usually leads to improvement. However, hepatic abscesses may occur and recurrent episodes of cholangitis can result in septic decompensation. Prompt referral of the patient to a transplant centre is therefore mandatory in cases with recurrent cholangitis, even in the absence of end-stage liver disease.

Inflammatory Bowel Disease

Medical management of PSC patients with active colitis is not different from that of other patients with ulcerative colitis. The common association of IBD with PSC requires annual screening colonoscopy, with particular attention paid to the right colon [9]. A recent study showed that among patients with concomitant IBD the 10- and 20-year risks for colorectal carcinoma were 14 and 31%, significantly higher than in patients without IBD [10]. Colonic resection prior to liver transplantation is usually required if colonic dysplasia or invasive cancer is present. Although this procedure carries the risk of liver decompensation with severely increased morbidity and mortality, delayed resection could result in tumour spread, especially under immunosuppressive therapy.

Pruritus

Pruritus is frequent and can be an extremely distressing symptom of PSC. Some patients even complain in the absence of jaundice. An increase in opioid receptors and the serotonin neurotransmitter system have been shown to be implicated [11, 12]. Treatments range from antihistamines, cholestyramine, ursodeoxycholic acid, rifampin, opiate antagonists and serotonin uptake inhibitors to plasmapheresis. In therapy resistant cases liver transplantation can be the only option.

Osteodystrophy

Osteoporosis and osteopenia is a common complication of cholestatic liver disease, and up to 80% of patients with primary biliary cirrhosis or PSC have an abnormal bone mass [13]. This can lead to vertebral body compression fractures resulting in severe, immobilizing back pain and atraumatic fractures of the skeleton. Therefore, bone mineral density measurement should be offered to these patients. Calcium and vitamin D supplementation should be instituted even though evidence showing an improvement in bone mass is lacking.

Cholangiocarcinoma

The risk of cholangiocarcinoma (CCA) is severely increased in patients with PSC, occurring in up to 20% of cases. There is evidence that biliary dysplasia is a precursor of CCA [14], but CCA in PSC patients remains difficult to diagnose. In a case-control study serum CA 19-9 appeared to have good ability to discriminate PSC patients with and without CCA [15]. Resection for CCA on a background of PSC is difficult because of parenchymal fibrosis and cirrhosis. In addition, even curative liver resection for CCA in PSC patients does not cure the underlying disease. However, early attempts 15 years ago to treat PSC patients with CCA by liver transplantation often failed because of recurrent cancer, which in most cases occurred during the first years after the operation. Within the past 5 years a new strategy has appeared, initiated by the Mayo Clinic (Rochester, Minn., USA) [16], for which the clinicians responsible reported a 5-year survival rate of 80% [17]. This protocol combines radiochemotherapy for 8–12 weeks prior to liver transplantation. Importantly, certain prerequisites are mandatory following this Mayo protocol. First, extrahepatic tumour growth must be excluded by extended pre-operative work up. Usually this requires PET-CT, fine needle aspiration of lymph nodes in the hepatoduodenal ligament, cholangiography and MRI. Second, tumour growth should be limited to 3 cm in radial diameter without the presence of intrahepatic metastases. In selected cases that fulfil all of these criteria living donor liver transplantation offers the only possibility to electively plan radiochemotherapy and liver transplantation.

Indication for Liver Transplantation

Besides the indication for liver transplantation for PSC patients with hilar CCA, end-stage liver disease due to PSC also requires liver transplantation. In general, the same criteria for liver transplantation can be applied in these cases as in other chronic liver diseases. However, unique circumstances may arise and require earlier evaluation for transplantation. Such circumstances include, recurrent bacterial cholangitis despite medical therapy, severe extrahepatic biliary obstruction which cannot be treated surgically, and refractory pruritus. The Model for End-Stage Liver Disease (MELD) score calculation, used for liver graft allocation in most countries, addresses neither the risk of sepsis due to recurrent cholangitis nor the psychological suffering due to untreatable pruritus. Therefore, additional points often have to be requested in PSC patients listed for liver transplantation.

Transplant Surgery

The transplantation procedure in PSC patients differs only in the connection of the bile duct. Since recipient bile duct tissue that is allowed to remain could result in fibrotic strictures or cancer development, the standard practice is to remove all bile duct tissue down to the pancreas head. Afterwards, the donor choledochus during implantation must be connected with the jejunum of the recipient (Roux-en-Y hepaticojejunostomy). The procedure is often more complicated because of previous biliary surgery or previous repeated bile drains. Although the optimal length of the jejunal limb has not been established, it is a common practice to use 40 cm of defunctionalized jejunum.

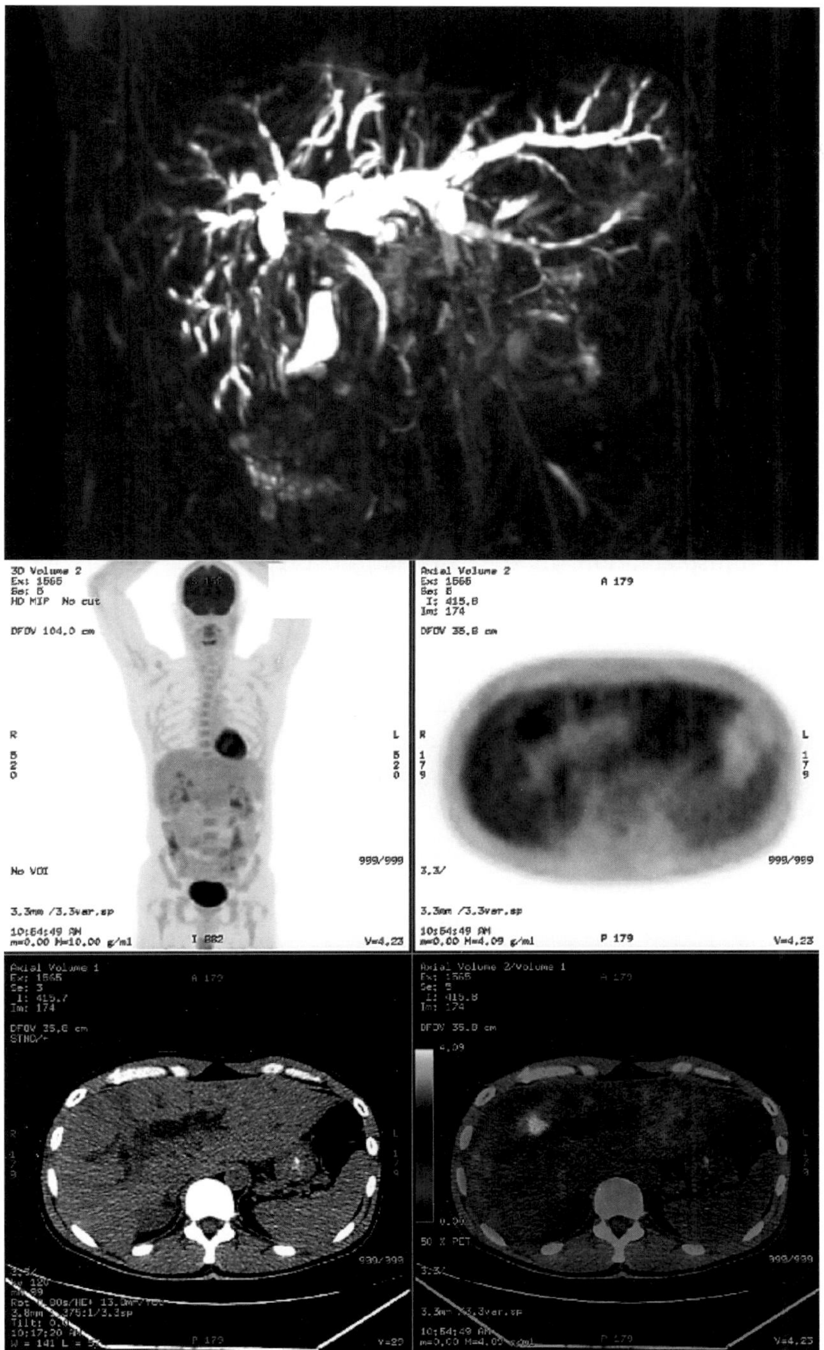

Fig. 1. MR and FDG-PET/CT from a 53-year-old male patient with primary sclerosing cholangitis and CCA. The MR cholangiogram shows both strictures and dilatation of the bile ducts. The FDG-PET/CT shows the tumour with increased glucose uptake.

Post-Transplantation Complications and Immunosuppressive Treatment

Patients undergoing liver transplantation for PSC have disease specific complications that may lead to additional morbidity and mortality.

Acute Cellular and Chronic Rejection
Several centres have identified an increased incidence of acute rejection in PSC patients after liver transplantation. Coexistent inflammatory bowel disease may also increase the risk of rejection. Most centres therefore prefer a tacrolimus-based immunosuppressive regimen, often augmented by mycophenolic acid and steroids. However chronic ductopenic rejection can occur with an incidence of up to 8–13%.

Inflammatory Bowel Disease following Liver Transplantation
The course of IBD following a liver transplantation is variable. While many patients have improvement in symptoms, some show severe worsening of IBD or even new onset of IBD after transplantation. In spite of that, no precipitating factors have been identified. For follow-up of ulcerative colitis an annual screening colonoscopy is recommended. In this context, a recent study [18] suggests that long-term use of calcineurin inhibitors may contribute to increased IBD activity. One hypothesis for this aggravation of IBD is that the mechanism of action of these calcineurin inhibitors is involved, as they are known to reduce the IL-2 dependent generation of regulatory T cells. These regulatory T cells have been recognized as being crucial for immunological homeostasis in the intestine. In addition, activation induced cell death of T cells is down regulated by calcineurin inhibitors. The increased risk of IBD in patients using tacrolimus in some studies could therefore well be explained by the fact that tacrolimus is a stronger calcineurin inhibitor than cyclosporine. In summary, risk factors for recurrence and de novo IBD after liver transplantation include currently:

- use of tacrolimus;
- active IBD at time of transplantation;
- short duration of IBD prior to transplantation.

Initiation of 5-aminosalicylate shortly after transplantation potentially reduces the risk for recurrence.

Post-Transplantation Survival

Candidates with PSC but without CCA have excellent survival if operated early enough. In a disease-specific analysis of the UNOS database, survival rates reached 90% at 1 and 84% at 2 years, these being higher than in the other patient subgroups [19]. Another study comparing recipient survival to that of comparable candidates without a transplant underlined the significant benefit of liver transplantation in PSC.

The adjusted mortality rate ratio for transplantation here was 0.31, which was 69% lower than that for patients remaining on the waiting list [20].

Recurrent Disease

Diagnosis of recurrence of PSC after liver transplantation can be challenging and is considered to be a diagnosis of exclusion. Chronic rejection, biliary obstruction and hepatic artery stenosis all can mimic recurrent PSC. Cholangiographically, the appearance of ischemic bile duct injury is no different from the classic appearance of PSC. Reported frequency of recurrence ranges from 9 to 47% [21]. Numerous studies have tried to identify risk factors for disease recurrence after liver transplantation [22–25]. A recent cohort study of 230 patients with a mean follow-up of over 6 years could show that colectomy was a significant factor for recurrent PSC. A colectomy before or during initial liver transplantation conferred a protective effect in subsequent grafts [26]. Another recent study showed that maintenance of steroid treatment for ulcerative colitis after transplantation was an independent risk factor [27]. Acute cellular rejection has also been associated with increased risk [28]. Whether recurrence of PSC adversely affects the outcome is unknown. Graft failure due to recurrence is rare and often confounded by presence of other forms of liver injury. Potential medical interventions include additional immunosuppressive agents and/or bile acid therapy.

Autoimmune Hepatitis

Issues before Transplantation

AIH is a chronic necro-inflammatory liver disease of unknown aetiology [29], with female predominance, and as many as 50% of patients are younger than 30 years of age. Clinical manifestations are highly variable and sometimes follow a fluctuating course. So far, no single clinical or biochemical test has been able to prove the presence of AIH. Diagnosis is based on characteristic histological, clinical, biochemical and serological findings. AIH shares common features with other autoimmune liver diseases such as primary biliary cirrhosis and PSC [30]. One clue to diagnosing AIH is the presence of other autoimmune diseases such as ulcerative colitis, thyroiditis or rheumatoid arthritis.

A mean annual incidence of 2–17 cases per 100,000 persons have been reported in Western European populations [31–33]. The prevalence of cirrhosis at presentation was comparable in patients older or younger than 60 years of age [34]. Long-term survival and average life expectancy is good in well-managed patients and is estimated to be comparable with those of the normal population [35]. On the other hand, untreated AIH has a high mortality and up to 40% of untreated patients with end-stage liver disease die within 6 months [36].

Underlying genetic predisposition is strongly suggested because of the association of the disease with certain human leucocyte antigens (HLA), particularly HLA B8, DR3, and DR4 [37]. Based on circulating antibodies, 2 types of AIH have been proposed. Type 1 AIH is the most common subclass, representing 80% of the cases. Antinuclear antibody, smooth muscle antibody, and perinuclear antineutrophil cytoplasmatic antibody are diagnostic autoantibodies in this subclass. Anti-soluble liver antigen/liver-pancreas antigen is sometimes helpful in the diagnosis. Type 2 AIH is characterized by anti-liver/kidney microsome-1 and anti-liver cytosol-1 autoantibodies occurring alone or together [29].

The International Autoimmune Hepatitis Group proposed revised diagnostic criteria based on 6 major studies in 1999 [37], and it presented a simplified scoring system for wider applicability in clinical practice in 2008 [38]. This score included autoantibodies, Ig G, histology and exclusion of viral hepatitis.

The clinical presentation does not differ from that of other forms of chronic hepatitis in most cases, ranging from asymptomatic disease to severe icteric hepatitis, and even fulminant hepatic failure [29].

The therapeutic goal of medical treatment should be complete normalization of transaminases with induction and maintenance of remission. Long-term immunosuppressive therapy with corticosteroids, usually in combination with azathioprine is considered to be the first-line medical treatment. Therapy for AIH is continued until remission is achieved. It has been proposed that treatment should not be withdrawn before 4 years of continued therapy [39]. Alternative treatment regimens have also been proposed, including cyclosporine, tacrolimus, mycophenolate mofetil and budesonide [40–43].

Indication for Liver Transplantation

In general, AIH responds well to anti-inflammatory or/and immunosuppressive treatment. Compared to PSC, transplantation in AIH should be delayed [44]. About 10% of patients intolerant or refractory to medical therapy and who develop decompensated cirrhosis may, however, require liver transplantation [45]. The transplant procedure for AIH does not differ from other end-stage liver disease.

Immunosuppressive Treatment

Because of the higher incidence of acute, chronic, and recurrent disease in the allograft, patients with AIH require aggressive immunosuppression. Withdrawal of these is rarely possible, and long-term corticosteroid treatment is often indispensable.

Acute rejection in patients transplanted for AIH is significantly higher than in those with alcoholic liver disease [46]. Acute rejection can occur in up to 80%, and of these 23–59% are steroid resistant and require treatment with ornithine-ketoacid transaminase. Within a cohort study younger patients who underwent grafting for AIH were at increased risk of chronic rejection. Additionally, among patients who developed chronic rejection, moderate/severe acute rejection was significantly more common [47]. The optimal immunosuppressive regimen has not been established yet. Most centres use a calcineurin inhibitor in combination with azathioprine or mycophenlate and corticosteroids.

Post-Transplantation Survival

The reported graft and patient survival rates at 5 years after liver transplantation are approximately 80 to 90% [29, 48]. In a single-centre experience, retrospective analysis showed a 3-year patient and graft survival of approximately 90%, which was the highest survival among all other indications for liver transplantation in their program [49].

Recurrent Disease

Currently there are no established diagnostic criteria for recurrent AIH. Diagnosis is based on increased concentration of transaminases or autoantibodies, hypergammaglobulinaemia, lobular or periportal hepatitis in the absence of viral hepatitis, and steroid dependency. A systematic review of 13 articles showed a recurrence rate of 22% for AIH [48]. A general approach in recurrent AIH is to increase corticosteroids, and some centres switch from cyclosporine to tacrolimus-based regimes. The usefulness of method is debatable since a systematic review by Gautam et al. [48] could show no statistically significant difference in recurrence between immunosuppression with cyclosporine or tacrolimus.

PSC and AIH: Similarities and Differences

While PSC affects more male patients and involves exclusively the intra- and extrahepatic biliary tree, AIH has a female prevalence of 78% and shows lymphoplasmacytic necroinflammatory infiltrates in periportal or paraseptal hepatic tissue. IBD is common in both types of autoimmune liver disorder but IBD is more frequent in PSC patients (table 1). The risk for CCA clearly increases in PSC patients while CCA in AIH has not been reported. Because of this risk of biliary cancer, liver transplantation is the treatment of choice even in early cases of PSC. In contrast, the treatment in

Table 1. Similarities and differences between AIH and PSC

	PSC	AIH
Patients	Young male	Young female
Initiating injury	Inflammation of extra and intrahepatic biliary tree	Periportal hepatitis
Transplantation		
Prior issues	IBD in 80%; risk for CCA	IBD in 20%
Indication	Early	Late
Surgery	Extended removal of biliary tract necessary	
Post-transplantation		
Acute rejection	69%	56–83%
Chronic rejection	8%	16%
10-year survival	70%	75%
Immnosuppressive treatment	Steroids + CNI + MMF	Steroids + CNI + MMF
Recurrent disease		
Incidence	PSC 11–37%	AIH 12–36%
Diagnosis	Histology and imaging of biliary tree	Serology, biochemistry, histology
Treatment	Ursodeoxycholic acid	Add or increase steroids
Risk factors	Acute cellular rejection; prolonged steroids; intact colon; ECD graft	HLA-DR3-positive recipient; necro-inflammation in native liver; lack of steroids
Graft loss	more frequent	less frequent

CNI = Calcineurin inhibitor; ECD = extended criteria donor; MMF = mycophenolate mofetil.

AIH consists primarily of medical regimen. Liver transplantation is considered only in AIH cases that develop decompensated liver disease.

From a technical point of view, liver transplantation in PSC patients always requires hepaticojejunostomy to completely remove the recipient's common bile duct. After transplantation, acute rejection periods are common in both PSC and AIH, and require therefore at least 2 immunosuppressive agents to maintain graft function. However, long-term survival is excellent and cumulates to 70–75% (table 1). On the other hand, recurrence of disease has been described in both disorders and is difficult to diagnose because histological findings are often non-specific due to other reasons, such as reperfusion injury, rejection, ischemia or sepsis. The diagnosis of PSC recurrence depends on the presence of typical cholangiographic findings (multiple non-anastomic biliary strictures) together with histologic alterations, after careful exclusion of all other causes (table 1). Recurrence of AIH is suspicious in the presence of typical autoantibodies, together with the presence of portal lymphatic plamacytic or lymphoplasmacytic infiltrates (table 1). Proposed risk factors for recurrent PSC include cold ischaemia time, number of cellular rejection episodes and cytomegalovirus infections. In addition, the absence of the colon after liver transplantation protects the patient from recurrent PSC. Moreover, extended donor criteria grafts appear to be significant risk factors for disease recurrence in PSC. Risk factors for recurrent AIH are the lack or absence of steroids after liver transplantation and high-grade inflammation in the native liver (table 1).

References

1 Papatheodoridis GV, Hamilton M, Rolles K, Burroughs AK: Liver transplantation and inflammatory bowel disease. J Hepatol 1998;28:1070–1076.

2 Czaja AJ, Davis GL, Ludwig J, Baggenstoss AH, Taswell HF: Autoimmune features as determinants of prognosis in steroid-treated chronic active hepatitis of uncertain etiology. Gastroenterology 1983;85:713–717.

3 Bambha K, Kim WR, Talwalkar J, Torgerson H, Benson JT, Therneau TM, Loftus EV Jr, et al: Incidence, clinical spectrum, and outcomes of primary sclerosing cholangitis in a United States community. Gastroenterology 2003;125:1364–1369.

4 Bernstein CN, Blanchard JF, Rawsthorne P, Yu N: The prevalence of extraintestinal diseases in inflammatory bowel disease: a population-based study. Am J Gastroenterol 2001;96:1116–1122.

5 Olsson R, Danielsson A, Jarnerot G, Lindstrom E, Loof L, Rolny P, Ryden BO, et al: Prevalence of primary sclerosing cholangitis in patients with ulcerative-colitis. Gastroenterology 1991;100:1319–1323.

6 Worns MA, Lohse AW, Neurath MF, Croxford A, Otto G, Kreft A, Galle PR, et al: Five cases of de novo inflammatory bowel disease after orthotopic liver transplantation. Am J Gastroenterol 2006;101:1931–1937.

7 Lindor KD, Kowdley KV, Luketic VA, Harrison ME, McCashland T, Befeler AS, Harnois D, et al: High-dose ursodeoxycholic acid for the treatment of primary sclerosing cholangitis. Hepatology 2009, E-pub ahead of print.

8 Pohl J, Ring A, Stremmel W, Stiehl A: The role of dominant stenoses in bacterial infections of bile ducts in primary sclerosing cholangitis. Eur J Gastroenterol Hepatol 2006;18:69–74.

9 Claessen MM, Lutgens MW, van Buuren HR, Oldenburg B, Stokkers PC, van der Woude CJ, Hommes DW, et al: More right-sided IBD-associated colorectal cancer in patients with primary sclerosing cholangitis. Inflamm Bowel Dis 2009, E-pub ahead of print.

10 Claessen MM, Vleggaar FP, Tytgat KM, Siersema PD, van Buuren HR: High lifetime risk of cancer in primary sclerosing cholangitis. J Hepatol 2009;50:158–164.

11 Bergasa NV, Jones EA: The pruritus of cholestasis: potential pathogenic and therapeutic implications of opioids. Gastroenterology 1995;108:1582–1588.

12 Richardson BP: Serotonin and nociception. Ann NY Acad Sci 1990;600:511–519; discussion 519–520.

13 Guichelaar MM, Kendall R, Malinchoc M, Hay JE: Bone mineral density before and after OLT: long-term follow-up and predictive factors. Liver Transpl 2006;12:1390–1402.

14 Fleming KA, Boberg KM, Glaumann H, Bergquist A, Smith D, Clausen OP: Biliary dysplasia as a marker of cholangiocarcinoma in primary sclerosing cholangitis. J Hepatol 2001;34:360–365.

15 Chalasani N, Baluyut A, Ismail A, Zaman A, Sood G, Ghalib R, McCashland TM, et al: Cholangiocarcinoma in patients with primary sclerosing cholangitis: a multicenter case-control study. Hepatology 2000;31:7–11.

16 Gores GJ, Nagorney DM, Rosen CB: Cholangiocarcinoma: is transplantation an option? For whom? J Hepatol 2007;47:455–459.
17 Rea DJ, Heimbach JK, Rosen CB, Haddock MG, Alberts SR, Kremers WK, Gores GJ, et al: Liver transplantation with neoadjuvant chemoradiation is more effective than resection for hilar cholangiocarcinoma. Ann Surg 2005;242:451–458; discussion 458–461.
18 Verdonk RC, Dijkstra G, Haagsma EB, Shostrom VK, Van den Berg AP, Kleibeuker JH, Langnas AN, et al: Inflammatory bowel disease after liver transplantation: risk factors for recurrence and de novo disease. Am J Transplant 2006;6:1422–1429.
19 Roberts MS, Angus DC, Bryce CL, Valenta Z, Weissfeld L: Survival after liver transplantation in the United States: a disease-specific analysis of the UNOS database. Liver Transpl 2004;10:886–897.
20 Merion RM, Schaubel DE, Dykstra DM, Freeman RB, Port FK, Wolfe RA: The survival benefit of liver transplantation. Am J Transplant 2005;5:307–313.
21 Kotlyar DS, Campbell MS, Reddy KR: Recurrence of diseases following orthotopic liver transplantation. Am J Gastroenterol 2006;101:1370–1378.
22 Vera A, Moledina S, Gunson B, Hubscher S, Mirza D, Olliff S, Neuberger J: Risk factors for recurrence of primary sclerosing cholangitis of liver allograft. Lancet 2002;360:1943–1944.
23 Goss JA, Shackleton CR, Farmer DG, Arnaout WS, Seu P, Markowitz JS, Martin P, et al: Orthotopic liver transplantation for primary sclerosing cholangitis: a 12-year single center experience. Ann Surg 1997; 225:472–481; discussion 481–473.
24 Gopal DV, Corless CL, Rabkin JM, Olyaei AJ, Rosen HR: Graft failure from severe recurrent primary sclerosing cholangitis following orthotopic liver transplantation. J Clin Gastroenterol 2003;37:344–347.
25 Kugelmas M, Spiegelman P, Osgood MJ, Young DA, Trotter JF, Steinberg T, Wachs ME, et al: Different immunosuppressive regimens and recurrence of primary sclerosing cholangitis after liver transplantation. Liver Transpl 2003;9:727–732.
26 Alabraba E, Nightingale P, Gunson B, Hubscher S, Olliff S, Mirza D, Neuberger J: A re-evaluation of the risk factors for the recurrence of primary sclerosing cholangitis in liver allografts. Liver Transpl 2009;15:330–340.
27 Cholongitas E, Shusang V, Papatheodoridis GV, Marelli L, Manousou P, Rolando N, Patch D, et al: Risk factors for recurrence of primary sclerosing cholangitis after liver transplantation. Liver Transpl 2008;14:138–143.
28 Alexander J, Lord JD, Yeh MM, Cuevas C, Bakthavatsalam R, Kowdley KV: Risk factors for recurrence of primary sclerosing cholangitis after liver transplantation. Liver Transpl 2008;14:245–251.
29 Krawitt EL: Autoimmune hepatitis. N Engl J Med 2006;354:54–66.
30 Lohse AW, zum Buschenfelde KH, Franz B, Kanzler S, Gerken G, Dienes HP: Characterization of the overlap syndrome of primary biliary cirrhosis (PBC) and autoimmune hepatitis: evidence for it being a hepatitic form of PBC in genetically susceptible individuals. Hepatology 1999;29:1078–1084.
31 Ritland S: The incidence of chronic active hepatitis in Norway: a retrospective study. Scand J Gastroenterol Suppl 1985;107:58–60.
32 Olsson R, Lindberg J, Weiland O, Nilsson L: Chronic active hepatitis in Sweden: the etiologic spectrum, clinical presentation, and laboratory profile. Scand J Gastroenterol 1988;23:463–470.
33 Hodges JR, Millward-Sadler GH, Wright R: Chronic active hepatitis: the spectrum of disease. Lancet 1982;1:550–552.
34 Al-Chalabi T, Boccato S, Portmann BC, McFarlane IG, Heneghan MA: Autoimmune hepatitis (AIH) in the elderly: a systematic retrospective analysis of a large group of consecutive patients with definite AIH followed at a tertiary referral centre. J Hepatol 2006;45:575–583.
35 Kanzler S, Lohr H, Gerken G, Galle PR, Lohse AW: Long-term management and prognosis of autoimmune hepatitis (AIH): a single center experience. Z Gastroenterol 2001;39:339–341; 344–338.
36 Soloway RD, Summerskill WH, Baggenstoss AH, Geall MG, Gitnick GL, Elveback IR, Schoenfield LJ: Clinical, biochemical, and histological remission of severe chronic active liver disease: a controlled study of treatments and early prognosis. Gastroenterology 1972;63:820–833.
37 Alvarez F, Berg PA, Bianchi FB, Bianchi L, Burroughs AK, Cancado EL, Chapman RW, et al: International Autoimmune Hepatitis Group Report: review of criteria for diagnosis of autoimmune hepatitis. J Hepatol 1999;31:929–938.
38 Hennes EM, Zeniya M, Czaja AJ, Pares A, Dalekos GN, Krawitt EL, Bittencourt PL, et al: Simplified criteria for the diagnosis of autoimmune hepatitis. Hepatology 2008;48:169–176.
39 Kanzler S, Gerken G, Lohr H, Galle PR, Meyer zum Buschenfelde KH, Lohse AW: Duration of immunosuppressive therapy in autoimmune hepatitis. J Hepatol 2001;34:354–355.
40 Malekzadeh R, Nasseri-Moghaddam S, Kaviani MJ, Taheri H, Kamalian N, Sotoudeh M: Cyclosporin A is a promising alternative to corticosteroids in autoimmune hepatitis. Dig Dis Sci 2001;46:1321–1327.
41 Aqel BA, Machicao V, Rosser B, Satyanarayana R, Harnois DM, Dickson RC: Efficacy of tacrolimus in the treatment of steroid refractory autoimmune hepatitis. J Clin Gastroenterol 2004;38:805–809.

42 Czaja AJ, Carpenter HA: Empiric therapy of autoimmune hepatitis with mycophenolate mofetil: comparison with conventional treatment for refractory disease. J Clin Gastroenterol 2005;39:819–825.
43 Csepregi A, Rocken C, Treiber G, Malfertheiner P: Budesonide induces complete remission in autoimmune hepatitis. World J Gastroenterol 2006;12:1362–1366.
44 Tillmann HL, Jackel E, Manns MP: Liver transplantation in autoimmune liver disease: selection of patients. Hepatogastroenterology 1999;46:3053–3059.
45 Reich DJ, Fiel I, Guarrera JV, Emre S, Guy SR, Schwartz ME, Miller CM, et al: Liver transplantation for autoimmune hepatitis. Hepatology 2000;32:693–700.
46 Hayashi M, Keeffe EB, Krams SM, Martinez OM, Ojogho ON, So SK, Garcia G, et al: Allograft rejection after liver transplantation for autoimmune liver diseases. Liver Transpl Surg 1998;4:208–214.
47 Milkiewicz P, Gunson B, Saksena S, Hathaway M, Hubscher SG, Elias E: Increased incidence of chronic rejection in adult patients transplanted for autoimmune hepatitis: assessment of risk factors. Transplantation 2000;70:477–480.
48 Gautam M, Cheruvattath R, Balan V: Recurrence of autoimmune liver disease after liver transplantation: a systematic review. Liver Transpl 2006;12:1813–1824.
49 Khalaf H, Mourad W, El-Sheikh Y, Abdo A, Helmy A, Medhat Y, Al-Sofayan M, et al: Liver transplantation for autoimmune hepatitis: a single-center experience. Transplant Proc 2007;39:1166–1170.

P.-A. Clavien
Department of Visceral- and Transplantation Surgery
University Hospital Zurich
Rämistrasse 100, CH–8091 Zürich (Switzerland)
Tel. +41 44 265 33 06, Fax +41 44 255 44 49, E-Mail clavien@chir.unizh.ch

Mayerle J, Tilg H (eds): Clinical Update on Inflammatory Disorders of the Gastrointestinal Tract.
Front Gastrointest Res. Basel, Karger, 2010, vol 26, pp 72–82

Hepatocellular Carcinoma

Markus Peck-Radosavljevic

Abteilung Gastroenterologie and Hepatologie, AKH and Medizinische Universität Wien, Vienna, Austria

Abstract

Hepatocellular carcinoma (HCC) is the sixth most common cancer worldwide. In most cases HCC is the consequence of chronic liver disease and in 80% of cases it is associated with liver cirrhosis. It is particularly prevalent in areas where chronic viral hepatitis is endemic, such as large parts of Africa and Asia. In its early stages, HCC can be treated quite effectively, so early diagnosis is key to successful therapy. However, since patients with early HCC are usually asymptomatic, identification and rigorous screening and surveillance of risk groups is essential. In most cases, a diagnosis can be made using non-invasive techniques. Treatment is stage dependent and best depicted by the Barcelona Clinic Liver Cancer (BCLC) classification. It subdivides treatment options into curative (BCLC stage 0 and A), palliative (BCLC stage B and C), and supportive (BCLC stage D). These include surgical, radiological, medical, and supportive interventions and offer very different outcomes. Great advances have been made in recent year regarding the identification of the optimal candidates for resection and liver transplantation, the use of radiofrequency ablation, the selection of candidates and the technical application of transarterial chemoembolization, and the introduction of truly effective medical therapies for HCC. The introduction of new medical treatments and the adjuvant or neoadjuvant use of these drugs are currently being explored in several large-scale trials globally. Significant advances in the management of HCC will be seen in the next years in all patients except BCLC stage D, where no drug will be able to improve outcome in the nearer future.

Incidence and Epidemiology

HCC is the sixth most common cause of cancer worldwide, with an annual incidence of over 711,000 new cases, and it is the third most common cause of cancer death, with an annual mortality of 679,000 patients [1].

Most commonly, HCC develops in cirrhosis, irrespective of the aetiology. In chronic hepatitis C it is estimated that about 20% of patients will develop cirrhosis eventually after 20–30 years of infection. For chronic hepatitis B, progression of liver disease depends on viral replication and the amount of liver injury incurred. The worldwide incidence of chronic hepatitis B is estimated at around 350 million cases, while chronic hepatitis C could affect as many as 170 million people worldwide.

Chronic hepatitis B is endemic in the developing world in Asia and Africa but is also prevalent in eastern Europe and other areas of the world. Infection in the developing countries occurs often times during birth or in the early childhood, sexual transmission is the main route of infection later in life. While nationwide vaccination programs were successful in stopping the spread of hepatitis B in the developed world, these vaccination programs will still take some time to exert an effect in developing countries, even if they are in place already.

In the West, chronic hepatitis C is today mostly contracted by intravenous drug abuse. Poor sanitary conditions, especially while delivering healthcare, are still common ways of contracting hepatitis C in developing countries.

Chronic viral hepatitis without cirrhosis is less commonly associated with the development of HCC. The yearly risk of developing HCC in patients with chronic hepatitis B without cirrhosis is reported to be around 0.5% in Asian series, with no corresponding data being available for the West. Whether chronic hepatitis C without cirrhosis can lead to HCC is not clear at the moment.

Once cirrhosis is established, the annual risk of developing HCC is estimated to be between 3 and 4% [2], largely irrespective of the aetiology of cirrhosis.

Screening and Surveillance

HCC in early stages is asymptomatic and only screening/surveillance can guarantee early detection. Several studies on the impact of surveillance on HCC detection have been conducted. A European study was able to show that improvement of HCC detection-rates over a 15-year time frame through more experience and better ultrasound machines resulted in smaller tumours at detection and improved survival. A randomized trial of HCC-surveillance with liver ultrasound and α-feto protein (AFP) every 6 months in 19,200 HBsAg+ patients in China demonstrated a 37% reduction in mortality after 5 years of surveillance, which was highly significant [3]. The resection rate was improved from 7.8% without surveillance to 46.5% in the surveillance group, even though the overall compliance with the surveillance program was only 58%. Surveillance by liver ultrasound every 6 months is now the accepted strategy applied by most centres, since shorter intervals do not offer additional benefit and longer intervals result in reduced surveillance benefit.

Diagnosis and Staging

According to recent guidelines, diagnosis of HCC can be established by non-invasive means in many cases, as long as the lesion is located within a cirrhotic liver [4]. Methods acceptable for diagnosis of HCC are contrast-enhanced ultrasound, triphasic multislice-CT or contrast-enhanced MRT. Lesions smaller than 1 cm should be followed with imaging every 3–6 months. Lesions 1–2 cm in diameter should show a typical HCC-pattern on 2 of the 3 imaging methods. If not, HCC should be confirmed by biopsy. Lesions larger than 2 cm need only 1 typical imaging result with 1 of the 3 imaging modalities mentioned above. Atypical lesions should be confirmed by biopsy, unless the AFP is above 200 ng/ml. Lesions in a non-cirrhotic liver require confirmation by biopsy in any case.

Staging of HCC can be performed according to various staging systems such as TNM, Okuda, CLIP and Cupi. The most widely used staging system today is the Barcelona Clinic Liver Cancer (BCLC) system and treatment algorithm [5]. It has the advantage of incorporating not only tumour-related factors into its decision tree but also those related to cirrhosis (such as the hepatovenous pressure gradient or the Child-Pugh stage) and general performance status (as assessed by the Eastern Cooperative Oncology Group performance-status).

Treatment

Initial classification in the BCLC-staging system into very different prognostic groups is done mostly on the basis of the severity of the underlying liver disease and general performance status, and only very crudely according to the tumour extension (fig. 1). Only at the next level does the BCLC-system take a more detailed look at the specific features of HCC, ultimately leading to 1 or 2 specific treatment options.

Treatment of HCC is divided into 3 major categories: curative treatments, palliative treatments, best supportive care without any specific HCC-therapy.

Curative treatment options include liver resection, orthotopic liver transplantation (OLT), and loco-ablative therapies such as radiofrequency ablation (RFA) and percutaneous ethanol injection (PEI). These treatments can offer 5-year survival rates of between 40 and 70% [6], and about 30% of patients in Western countries will fall into this category.

The palliative treatment options available are transarterial chemoembolization (TACE) and, to a lesser extent, transarterial embolization. Treatment with sorafenib, the first systemic treatment ever to show efficacy against HCC in a prospective randomized controlled trial, is the other palliative treatment option. Median survival with palliative treatments ranges between 11 and 20 months [6], and about 50% of Western patients will require palliative therapy. Best supportive care offers a medial survival of less than 3 months and is the only option available to about 20% in Western countries [6].

Patients in the developing world with a higher percentage of hepatitis B associated HCC and with less rigorous HCC-screening and surveillance programs will likely have a higher fraction of patients only amenable to palliative and supportive care.

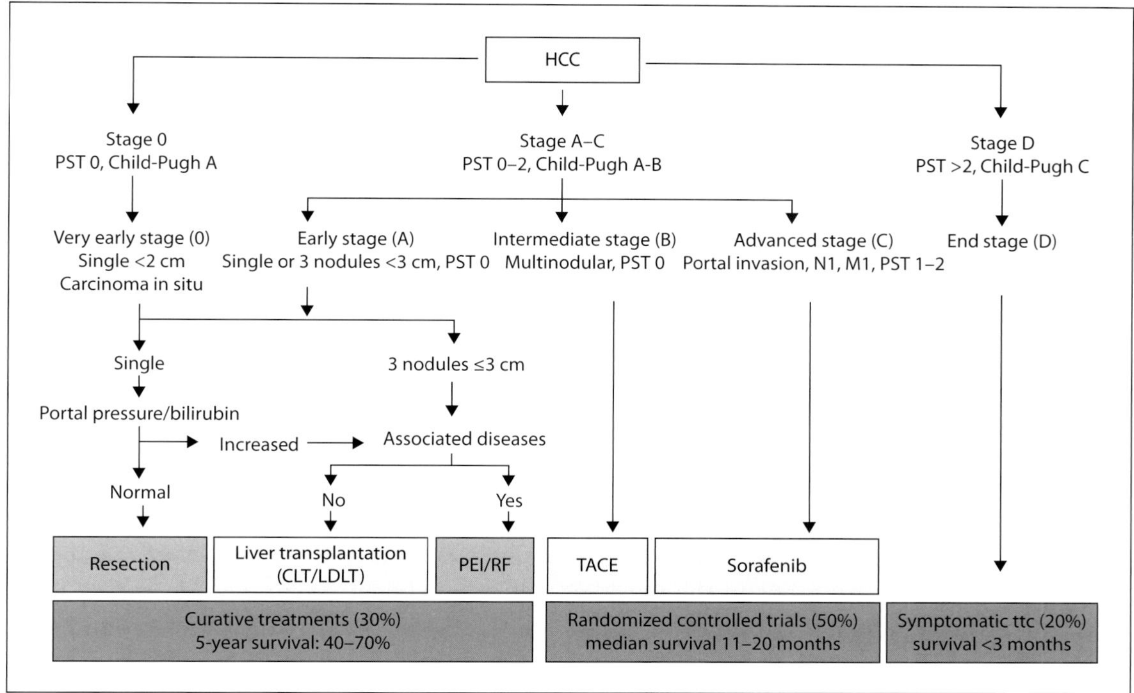

Fig. 1. The BCLC-staging and treatment algorithm of HCC in its current form [6]. PST = ECOG Performance Status Test.

Surgical Interventions for Advanced HCC

Surgical treatments as well as local ablative therapies are termed curative treatments for HCC but can only be applied in selected patients (fig. 1).

The surgical interventions to treat HCC are either liver resection or OLT. Neither are suitable to treat advanced HCC.

Resection is a very good option in non-cirrhotic patients, which amount to 5–10% of patients in the West and up to 40% of patients in Asia [2]. In non-cirrhotic patients, resection can also be employed for the treatment of more advanced HCC, as long as there is enough functional reserve left in the remnant liver. Surgical resection has been much aided by advances in the development of surgical equipment like ultrasonic dissectors, which obviate the need for a Pringle manoeuvre during resection [7].

If the feeling is that there would not be enough functional liver reserve left after large resections, portal vein embolization of the part of the liver to be resected can be used to induce hypertrophy of the remnant liver prior to resection using histoacryl glue [7].

Liver resection in patients with cirrhosis is a more difficult task and can only be performed in 5–10% of cirrhotics with HCC. Several factors have been proposed to triage patients for resection: only compensated patients with Child-Pugh A cirrhosis should be considered. In addition, it is quite common to assess the plasma retention rate of indocyanine at 15 min (indocyanine green clearance at 15 min), which should amount to less than 20% [7]. Absence of clinically significant portal hypertension has been shown to be a very reliable predictor for prevention of decompensation or death after resection in well-designed trials [8]; it is routinely performed in our institution before resection and has very good discriminative power. Patients with an hepatovenous pressure gradient less than 10 mmHg and normal bilirubin do best, while both an increase in portal pressure as well as an elevated bilirubin lead to increased complications after resection, as well as patient death [9].

As for technical advances in liver resection, techniques like portal vein embolization to augment the remnant liver, intermittent hepatic artery inflow obstruction instead of the classical Pringle-manoeuvre to minimize blood loss without causing major ischemia-reperfusion injury to the liver, and resection of lesions including a tumour thrombus in the portal as well as the hepatic vein have decreased perioperative mortality to less than 1% and can offer 5-year survival of around 50% [10].

OLT is the only really curative treatment option if applied in well-selected patients according to criteria termed the 'Milan criteria' [11]. The major advantage of OLT is that it cures not only the tumour but also the underlying cirrhosis and often also the underlying liver disease, except for patients with viral hepatitis, where recurrence after OLT is universal. While today recurrence of hepatitis B can be controlled quite successfully in most cases, treatment and control of chronic hepatitis C after OLT is more difficult and not attainable in all instances.

Recently, attempts have been made to push the size limits of the Milan-criteria to allow OLT in more patients with HCC [12]. These 'UCSF criteria' allow transplantation of patients with a single tumour up to 6.5 cm in diameter and up to 3 tumours <3 cm with a total tumour diameter not exceeding 8 cm, and seem to give outcomes comparable to the Milan criteria with regard to recurrence-free and overall patient survival, even with prospective radiologic imaging [12]. Unfortunately, in about 30% of cases, using pre-transplant radiologic imaging still results in under-staging of the tumour extension in patients exceeding the Milan but within the UCSF criteria. In addition, those patients have significant higher recurrence rates and poorer survival compared to correctly staged patients [12]. In centres following the UCSF criteria, an additional 20% with HCC will be able to undergo transplant and, with that, a potentially curative treatment. The recently proposed 'Up to 7 criteria' give even more flexibility and allow transplantation of patients with multiple small tumours [13]. Since microvascular invasion was a significant predictor of recurrence and survival in this large retrospective analysis, calculation of the true recurrence risk using the Up to 7 criteria would require a tumour biopsy, which is not standard of care in all transplant-

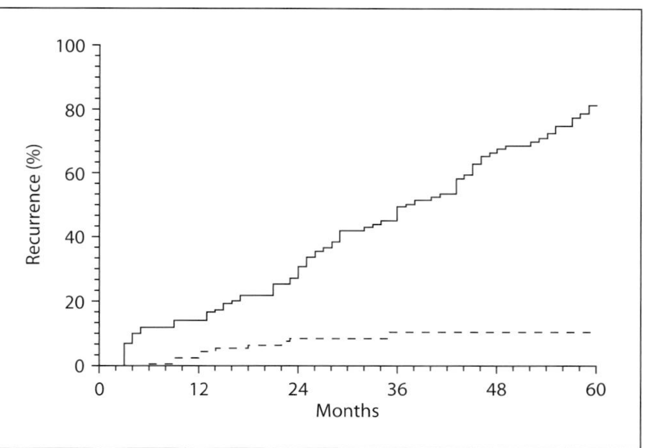

Fig. 2. Recurrence after RFA is mostly due to de novo tumour formation, not local recurrence [15]. HCC recurrence in patients with early-stage HCC who received RFA. Rates of recurrence within the treated tumor (i.e. local tumour progression; dotted line) were 4% at 1 year, 10% at 3 years, and 10% at 5 years. Rates of recurrence with new tumours (continuous line) were 14% at 1 year, 49% at 3 years, and 81% at 5 years.

centres. Those centres that want to be on the safe side with minimum tumour recurrence after OLT will currently stick to the Milan-criteria, since those will give a wider safety-margin for radiologic under-staging of HCC. In the setting of living donor transplantation, expansion of the criteria seems to be acceptable if the patients are rightly informed about the potential dangers of early tumour recurrence in case of radiologic under-staging.

Locoablative Therapies with Curative Intent

Excellent 5-year survival rates for local ablative therapies like RFA were reported in selected patients [14, 15]. Patients with single tumours and Child A cirrhosis excluded from surgery for various reasons had 5-year survival rates around 61% [15]. In an even more select patient-group of Child-A patients with single HCCs and a maximum tumour diameter of 2 cm, 5-year survival rates were even more impressive, at over 68% [14]. In a prospective randomized trial of 180 patients with HCC <5 cm, the equivalence of RFA to surgical resection could be documented for the whole study population as well as the subgroups with HCCs <3 cm and those with tumours between 3.1 and 5 cm in diameter [16]. Five-year tumour recurrence rates of RFA were around 80% in both Italian studies, with local recurrence being responsible for only 3% [14] to 10% [15] of recurrence. RFA is listed as a potentially curative treatment since most tumour recurrences are due to de-novo formation of a secondary tumour in the remnant cirrhotic liver (fig. 2). Strategies to reduce these recurrence rates are urgently needed.

PEI was the predecessor of RFA; it can be easily applied by ultrasound or CT/MR guidance, is cheap, and safe. PEI was compared to RFA in several prospective randomized trials and, even though the quality of some of these studies with regard to the PEI-technique could be questioned, was repeatedly shown to be inferior to RFA both with regards to local tumour control and overall patient survival [17].

PEI is mostly reserved today for lesions not amenable to RFA: those close to major blood vessels and large bile ducts, lesions close to the liver surface, and lesions in hard to access locations for the fairly large bore RFA-probes, which still might be reachable by the thin PEI-needles.

Established palliative Therapeutic Intervention for Advanced HCC

For intermediate (BCLC stage B) and advanced stage (BCLC stage C) patients, no curative treatment options are available to date. These patients can be treated with palliative treatment option: TACE and systemic treatments (currently only sorafenib).

Transarterial Chemoembolization

TACE is the standard treatment for patients with intermediate (BCLC stage B) disease, and in meta-analysis it showed a 2-year survival of 41 vs. 27% in control patients [18]. In this meta-analysis, the only 2 individual trials showed a survival benefit for TACE at 2 years, while 4 trials did not show a significant effect. Overall response rate to TACE was 35%. Mean survival in a randomized prospective controlled trial of TACE in mostly hepatitis C associated HCC was 28.7 months in the TACE group versus 17.9 months in the control group [19]. In a very large cohort of over 8,000 TACE-patients followed prospectively in Japan, 88% of which had virus-induced HCC, the median survival in treated patients was 34 months with a 5-year survival of 26% [20]. TACE by itself is not a well-defined technique. It is performed very differently with regards to the chemotherapeutic agent used (mostly doxorubicin or cisplatin), the use of lipiodol, as well as the use of gelfoam particles of different size or beadblock [4]. Only recently, a pilot trial of drug-eluting beads of defined size and loaded with a defined amount of doxorubicin was able to show an increased tumour response rate with a reduced systemic doxorubicin exposure [21]. Further development of this and other TACE-devices will allow more homogenous TACE-procedures with more comparable performance characteristics.

Currently, strategies combining different treatment modalities are being explored. The combination of drug-eluting beads with RFA was tested in a pilot trial [22]. Patients with a median tumour diameter of 5 cm and incomplete response to RFA were subsequently treated with TACE, which increased the tumour necrosis area by 60%, leading to complete response in 60% and to residual tumour of <10% of target lesion area in 30% of patients. Combination treatment shows promise for treatment of more advanced and complex lesions and could allow longer stabilization of disease.

Drug Treatment for Advanced HCC

For patients with BCLC stage C hepatocellular carcinomas, no effective treatments were available for many years despite continuous efforts in the development of chemotherapeutic drugs [23]. Meta-analyses, however difficult to perform, were unable to demonstrate a survival benefit for doxorubicin or 5-FU [24]. Uncontrolled studies and poorly defined study populations are major problems in assessment in many of the studies of systemic therapy for advanced stage HCC and are partly responsible for the scarcity of relevant data in this field.

In a large, well conducted randomized, placebo-controlled trial in 2007, the multikinase inhibitor sorafenib became the first systemic therapy to be shown to have a survival benefit in advanced stage HCC [25]. Sorafenib is a targeted agent that has anti-angiogenic and antiproliferative actions, inhibiting vascular endothelial growth factor receptor and platelet-derived growth factor receptor signalling and the RAF-MEK-ERK pathway. In the SHARP trial, sorafenib was able to significantly improve the overall survival from 7.9 months to 10.7 months and to improve the time to radiologic progression from 2.8 months to 5.5 months compared to placebo. Patient groups were well balanced with 82% having a BCLC stage C tumour and 70% having extrahepatic disease or vascular invasion. The treatment effects were very consistent among the different patient subgroups. Since 88% of patients were recruited in Europe, only about 47% of patients had underlying liver disease of viral aetiology, 28% chronic hepatitis C and 19% chronic hepatitis B [25]. Subgroup analysis of the hepatitis C patients from the SHARP-trial showed an overall survival of 14 months in the treatment groups compared to 7.9 months in the HCV-positive placebo patients. Time to progression was improved from 2.8 to 7.6 months, being highly significant for both analysis. Efficacy of sorafenib in the patients with chronic hepatitis C is consistent with the whole study population and side effects are also comparable.

A parallel study to the SHARP trial was conducted in the Asia-Pacific region, which was terminated early after enrolling 226 patients when the results of the SHARP trial became available [26]. In the Asia-Pacific trial, over 80% of the patients had chronic viral hepatitis (over 70% chronic hepatitis B), more patients had multilocular disease, and 50% of patients had lung metastasis. Again, sorafenib was able to show a highly significant improvement of survival over the control (6.2 vs. 4.1 months) and in time to progression (2.8 vs. 1.4 months). Most likely due to the different patient and disease characteristics, the absolute survival and time to progression figures were much smaller than in the European trial, highlighting the effects of the underlying liver disease on outcome in HCC-patients [6].

Non-Established and Experimental Treatment Options

Development of novel strategies for the treatment of HCC is a very active field of research.

Currently, there are a large number of ongoing clinical trials on new medical treatments, which mostly explore the application of more multikinase inhibitors with modes of action that are similar to or complement that of sorafenib. These include sunitinib [27, 28] and brivanib [29], which have entered phase III development, as well as erlotinib, cetuximab, bevacizumab and many others. Likewise, drugs with different modes of action, such as mTOR-inhibitors or thalidomide, are being explored in trials.

Radiation therapy is another area of active investigation. The concept of systemic internal radiation therapywas explored some time ago and is now being reintroduced into clinical development using ^{90}Y-labelled microspheres. In a recent phase II study it showed a favourable toxicity and efficacy profile in patients with advanced HCC [30]. It seems to be particularly interesting for patients with portal vein thrombosis, where other strategies, such as TACE, cannot be employed. The same holds true for external radiation therapy using stereotactic radiation, where median survival in patients not amenable to standard treatments such as TACE was 11.7 months [31], which is very similar to the survival of patients treated with sorafenib in the SHARP trial. The role of radiation treatment of HCC is to be established in future trials, comparing radiation therapy to locoregional therapies or to systemic treatments such as sorafenib. In addition, exploration of combination treatments of systemic agents with radiation therapy could be of interest.

Surgical and therapeutic interventions in HCC have come of age and the current treatment algorithms, even though not universally established by grade A evidence from clinical trials, are largely based on sound scientific rationale.

References

1 Garcia M, Jemal A, Ward EM, Center MM, Hao Y, Siegel RL, Thun MJ: Global Cancer Facts & Figures 2007. Atlanta, American Cancer Society, 2007. www.cancer.org.

2 Llovet JM, Bruix J: Novel advancements in the management of hepatocellular carcinoma in 2008. J Hepatol 2008;48(suppl 1):S20–S37.

3 Zhang BH, Yang BH, Tang ZY: Randomized controlled trial of screening for hepatocellular carcinoma. J Cancer Res Clin Oncol 2004;130:417–422.

4 Bruix J, Sherman M: Management of hepatocellular carcinoma. Hepatology 2005;42:1208–1236.

5 Llovet JM, Bru C, Bruix J: Prognosis of hepatocellular carcinoma: the BCLC staging classification. Semin Liver Dis 1999;19:329–338.

6 Llovet JM, Di Bisceglie AM, Bruix J, Kramer BS, Lencioni R, Zhu AX, Sherman M, et al: Design and endpoints of clinical trials in hepatocellular carcinoma. J Natl Cancer Inst 2008;100:698–711.

7 Marin-Hargreaves G, Azoulay D, Bismuth H: Hepatocellular carcinoma: surgical indications and results. Crit Rev Oncol Hematol 2003;47:13–27.

8 Bruix J, Castells A, Bosch J, Feu F, Fuster J, Garcia Pagan JC, Visa J, et al: Surgical resection of hepatocellular carcinoma in cirrhotic patients: prognostic value of preoperative portal pressure. Gastroenterology 1996;111:1018–1022.

9 Llovet JM, Fuster J, Bruix J: Intention-to-treat analysis of surgical treatment for early hepatocellular carcinoma: resection versus transplantation. Hepatology 1999;30:1434–1440.

10 Makuuchi M, Sano K: The surgical approach to HCC: our progress and results in Japan. Liver Transpl 2004;10:S46–S52.

11 Mazzaferro V, Regalia E, Doci R, Andreola S, Pulvirenti A, Bozzetti F, Montalto F, et al: Liver transplantation for the treatment of small hepatocellular carcinomas in patients with cirrhosis. N Engl J Med 1996;334:693–699.

12 Yao FY, Xiao L, Bass NM, Kerlan R, Ascher NL, Roberts JP: Liver Transplantation for hepatocellular carcinoma: validation of the UCSF-expanded criteria based on preoperative imaging. Am J Transplant 2007;7:2587–2596.

13 Mazzaferro V, Llovet JM, Miceli R, Bhoori S, Schiavo M, Mariani L, Camerini T, et al: Predicting survival after liver transplantation in patients with hepatocellular carcinoma beyond the Milan criteria: a retrospective, exploratory analysis. Lancet Oncol 2009; 10:35–43.

14 Livraghi T, Meloni F, Di Stasi M, Rolle E, Solbiati L, Tinelli C, Rossi S: Sustained complete response and complications rates after radiofrequency ablation of very early hepatocellular carcinoma in cirrhosis: is resection still the treatment of choice? Hepatology 2008;47:82–89.

15 Lencioni R, Cioni D, Crocetti L, Franchini C, Pina CD, Lera J, Bartolozzi C: Early-stage hepatocellular carcinoma in patients with cirrhosis: long-term results of percutaneous image-guided radiofrequency ablation. Radiology 2005;234:961–967.

16 Chen MS, Li JQ, Zheng Y, Guo RP, Liang HH, Zhang YQ, Lin XJ, et al: A prospective randomized trial comparing percutaneous local ablative therapy and partial hepatectomy for small hepatocellular carcinoma. Ann Surg 2006;243:321–328.

17 Shiina S, Teratani T, Obi S, Sato S, Tateishi R, Fujishima T, Ishikawa T, et al: A randomized controlled trial of radiofrequency ablation with ethanol injection for small hepatocellular carcinoma. Gastroenterology 2005;129:122–130.

18 Llovet JM, Bruix J: Systematic review of randomized trials for unresectable hepatocellular carcinoma: chemoembolization improves survival. Hepatology 2003;37:429–442.

19 Llovet JM, Real MI, Montana X, Planas R, Coll S, Aponte J, Ayuso C, et al: Arterial embolisation or chemoembolisation versus symptomatic treatment in patients with unresectable hepatocellular carcinoma: a randomised controlled trial. Lancet 2002; 359:1734–1739.

20 Takayasu K, Arii S, Ikai I, Omata M, Okita K, Ichida T, Matsuyama Y, et al: Prospective cohort study of transarterial chemoembolization for unresectable hepatocellular carcinoma in 8510 patients. Gastroenterology 2006;131:461–469.

21 Varela M, Real MI, Burrel M, Forner A, Sala M, Brunet M, Ayuso C, et al: Chemoembolization of hepatocellular carcinoma with drug eluting beads: efficacy and doxorubicin pharmacokinetics. J Hepatol 2007;46:474–481.

22 Lencioni R, Crocetti L, Petruzzi P, Vignali C, Bozzi E, Della Pina C, Bargellini I, et al: Doxorubicin-eluting bead-enhanced radiofrequency ablation of hepatocellular carcinoma: a pilot clinical study. J Hepatol 2008;49:217–222.

23 Thomas MB, O'Beirne JP, Furuse J, Chan AT, Abou-Alfa G, Johnson P: Systemic therapy for hepatocellular carcinoma: cytotoxic chemotherapy, targeted therapy and immunotherapy. Ann Surg Oncol 2008; 15:1008–1014.

24 Mathurin P, Rixe O, Carbonell N, Bernard B, Cluzel P, Bellin MF, Khayat D, et al: Review article: overview of medical treatments in unresectable hepatocellular carcinoma: an impossible meta-analysis? Aliment Pharmacol Ther 1998;12:111–126.

25 Llovet JM, Ricci S, Mazzaferro V, Hilgard P, Gane E, Blanc JF, de Oliveira AC, et al: Sorafenib in advanced hepatocellular carcinoma. N Engl J Med 2008;359: 378–390.

26 Cheng AL, Kang YK, Chen Z, Tsao CJ, Qin S, Kim JS, Luo R, et al: Efficacy and safety of sorafenib in patients in the Asia-Pacific region with advanced hepatocellular carcinoma: a phase III randomised, double-blind, placebo-controlled trial. Lancet Oncol 2009;10:25–34.

27 Zhu AX, Sahani DV, di Tomaso E, Duda D, Sindhwani V, Yoon SS, Blaszkowsky LS, et al: A phase II study of sunitinib in patients with advanced hepatocellular carcinoma. J Clin Oncol 2007;25 (suppl):A4637.

28 Faivre SJ, Raymond E, Douillard J, Boucher E, Lim HY, Kim JS, Lanzalone S, et al: Phase II trial investigating the efficacy and safety of sunitinib in patients with unresectable hepatocellular carcinoma (HCC). Eur J Cancer Supplements 2007;5:270.

29 Raoul J-L, Finn RS, Kang Y-K, Park J-W, Harris R, Coric V, Baudelet C, et al: An open-label phase II study of first- and second-line treatment with brivanib in patients with hepatocellular carcinoma (HCC). J Clin Oncol 2009;27(suppl 1):A4577.

30 Kulik LM, Carr BI, Mulcahy MF, Lewandowski RJ, Atassi B, Ryu RK, Sato KT, et al: Safety and efficacy of 90Y radiotherapy for hepatocellular carcinoma with and without portal vein thrombosis. Hepatology 2008;47:71–81.

31 Tse RV, Hawkins M, Lockwood G, Kim JJ, Cummings B, Knox J, Sherman M, et al: Phase I study of individualized stereotactic body radiotherapy for hepatocellular carcinoma and intrahepatic cholangiocarcinoma. J Clin Oncol 2008;26:657–664.

Markus Peck-Radosavljevic, MD
Abteilung Gastroenterologie and Hepatologie, Klinik Innere Medizin III
AKH und Medizinische Universität Wien
Währinger Gürtel 18–20, AT–1090 Vienna (Austria)
Tel. +43 1 40400 4741, Fax +43 1 40400 4735, E-Mail markus.peck@meduniwien.ac.at

Mayerle J, Tilg H (eds): Clinical Update on Inflammatory Disorders of the Gastrointestinal Tract.
Front Gastrointest Res. Basel, Karger, 2010, vol 26, pp 83–94

Coeliac Disease

Detlef Schuppan · Yvonne Junkler

Division of Gastroenterology, Beth Israel Deaconess Medical Center, Harvard Medical School, Boston, Mass., USA

Abstract

Coeliac disease (c.d.) is an inflammatory disorder of the small intestine that affects >0.5% of Western, Middle Eastern and North African populations. Disease manifestations range from asymptomatic to severe malabsorption, autoimmune disorders and (intestinal) lymphoma. A strict gluten-free diet is usually curative, but elderly patients can develop refractory c.d. or intestinal T cell lymphoma. c.d. shares immunological features with Crohn's disease, but it also displays unique features: (1) defined ingested gluten proteins from wheat and related cereals trigger a T cell-mediated adaptive immune response; (2) patients display serum autoantibodies to the ubiquitous enzyme tissue transglutaminase (TG2), and (3) gluten peptide presentation occurs by HLA-DQ2 or -DQ8. TG2 is instrumental in c.d. pathogenesis, since it deamidates certain gluten peptides which increases the peptides' affinity to HLA-DQ2 or -DQ8, resulting in a potentiated CD4+ T helper 1 T cell activation, which is a major driving force of villus atrophy and crypt hyperplasia, the hallmarks of c.d. Cereal proteins can also trigger an innate immune response that is directed at the intestinal epithelium and especially at dendritic cells and macrophages and which acts in concert with adaptive immunity. Non-dietary therapies are needed for patients with high gluten sensitivity and refractory c.d and much desired by patients with classical c.d. Some therapies are currently entering phase I or II clinical trials. Strategies include: (1) (genetically) modified cereals; (2) enzymatic destruction of immunogenic gliadin peptides; (3) blockage of intestinal permeability; (4) inhibition of intestinal TG2; (5) use of HLA-DQ2 blockers; (6) inhibition of proinflammatory cytokines, and (7) induction of tolerance to oral gluten.

Coeliac Disease, Epidemiology and Symptoms

Coeliac disease (c.d.) is an inflammatory small intestinal disorder which can lead to global malabsorption [1–4]. c.d. is triggered by the storage proteins (gluten) of wheat, barley and rye. These induce an inflammatory response in genetically predisposed individuals that results in intestinal crypt hyperplasia and villus atrophy. A life-long strict gluten-free diet is the only available treatment. However, this diet is difficult to maintain and often excludes patients from a normal social life.

While c.d. was considered a rare disease in the past, the advent of modern serological screening tests (autoantibodies to tissue transglutaminase or endomysium)

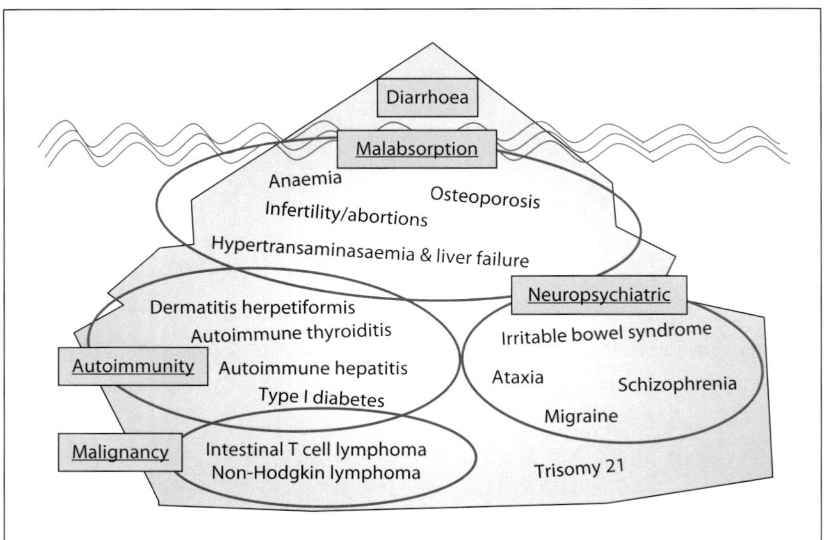

Fig. 1. The coeliac iceberg and diseases that are associated with c.d. Only a minority of patients presents with classical symptoms such as diarrhoea and malabsorption. Numerous other pathologies can be associated with c.d. and can be the major presenting feature (atypical c.d.). Any of these symptoms justifies autoantibody screening and, if positive, confirmation by duodenal biopsies. Some of the listed disorders, especially the neuropsychiatric manifestations, show only (unspecific) antibodies to gluten, but frequently lack autoantibodies to TG2 and intestinal lesions, ruling out conventional c.d. This has lead to the term 'gluten sensitivity' which may nonetheless be due to innate immune responses to cereal components.

followed by intestinal biopsy confirmation revealed prevalences in most Western, Middle Eastern and North African countries ranging from 1:80–1:200 [1–6]. Recent data from Finland even indicate a prevalence of >1:50 in subjects >52 years of age [7]. Most (>80%) of these screening-detected c.d. patients show no, minor or non-diarrhoea associated clinical symptoms (clinically silent, oligosymptomatic or atypical c.d.). The range of c.d. includes anaemia, osteoporosis and an often compromised wellbeing and quality of life, frequently without concomitant diarrhoea [5–7]. Atypical c.d. can present with otherwise unexplained neurological disorders such as peripheral neuropathy schizophrenia, autism, ataxia, or with aphthous stomatitis, arthritis, infertility, hypertransaminasaemia, and even liver failure [1–7]. In addition, c.d. is frequently found in conjunction with autoimmune diseases, such as type 1 diabetes, autoimmune thyroiditis, autoimmune hepatitis, dermatitis herpetiformis or autoimmune alopecia [1–8] (fig. 1). Lastly, patients with long-standing undetected and untreated symptomatic c.d. are at a increased risk to develop enteropathy associated T cell lymphoma, small bowel adenocarcinoma and other cancers of the gastrointestinal tract [9–10].

Pathogenesis of Coeliac Disease

Gluten, the Trigger

Some knowledge about the biochemistry of cereal proteins is necessary to understand c.d. pathogenesis. Gluten is the storage protein of wheat which exists in ancient (diploid or tetraploid), as well as in modern (hexaploid) variants. Each wheat variant produces >50 gliadins, which are structurally related proteins containing ~250–500 amino acids, and a lower number of high and low molecular weight glutenins containing ~650–800 and ~270–320 amino acids, respectively. It is mainly the ethanol-insoluble glutenins which are responsible for the desired baking properties. However, an admixture of the alcohol-soluble gliadins is also needed. Both gliadins and glutenins display a high content of the amino acids glutamine (32–56%) and proline (15–30%), and due to their cysteine content, glutenins can form complex homopolymers and heteropolymers with gliadins [11]. Storage proteins similar to gliadins (generally termed prolamines) are found in rye (secalins) and barley (hordeins) which can also cause c.d., while the avenins of oats and, for example, the zeins of rice are more distantly related and do not cause the disease. On the basis of their electrophoretic properties and primary structure the gliadins are subdivided into the classes of α-, γ and ω-gliadins, which can be further resolved into distinct proteins, such as α1–11, γ, and ω1–5.

Antigen Presentation by HLA-DQ2 or -DQ8

Essentially all patients with c.d. share the heterodimeric human lymphocyte antigen class II genes HLA-DQ2 (or -DQ8) as a necessary but not sufficient genetic background. HLA-DQ2 (DQ8) are expressed on antigen-presenting cells, such as macrophages and dendritic cells but also on B cells. Gluten peptides, with their unique sequences that are rich in proline, glutamine and certain hydrophobic amino acids, are preferentially presented by these c.d.-associated HLA class II molecules, which triggers the activation of gluten-specific CD4+ T helper 1 (Th1) cells in the mesenteric lymph nodes, followed by their homing back to the intestinal lamina propria. These activated Th1 T cells are instrumental in inducing architectural mucosal changes that result in the observed crypt hyperplasia and villus atrophy [3, 4,11]. Proposed mechanisms include the release of proinflammatory cytokines such as IFN-γ that induce matrix metalloproteiase release and activation from intestinal fibroblasts [12].

Another hallmark of c.d. is that patients develop mucosal (immunoglobulin A) autoantibodies to the enzyme tissue transglutaminase (transglutaminase 2, TG2) [13]. TG2 is expressed by many cell types (mainly by endothelial cells, fibroblasts and inflammatory cells) and is usually stored intracellularly [14]. In tissues that are mechanically stressed or during inflammation, TG2 is secreted and associates with the extracellular matrix where it binds to reticulin fibres and where it can become activated due to high ambient calcium concentrations. TG2 then reacts with certain glutamine residues of a few selected extra- and intracellular proteins, thereby covalently tethering these proteins to a lysine residue of a second protein, which results in irreversible crosslink formation. However, in the absence of a donor protein and at slightly acidic pH, as prevails during inflammation, TG2 can also merely deamidate the substrate glutamines of the donor protein, generating a protein or peptide with a negatively charged glutamic acid residue. Since gluten proteins have a high content in glutamine and proline residues, they are excellent substrates for crosslinking and deamidation by TG2 [15, 16]. Thus a set of de novo generated, negatively charged gluten peptides can bind more strongly to HLA-DQ2 (or -DQ8) than to their non-deamidated parent molecules, and this higher affinity leads to a more rigorous gluten-specific CD4+ Th1 T cell activation. More than 50 distinct consensus sequences for TG2-mediated deamidation have been identified in gluten and the related proteins from barley and rye [16–22]. Importantly, these peptides are unusually resistant to digestion by gastrointestinal proteases such as pepsin, trypsin and chymotrypsin, which explains their availability and activity in the (proximal) small intestine [19, 23].

Innate Immunity to Gluten

Components from wheat, rye or barley can also trigger innate immune responses in intestinal epithelial cells as well as in monocytes, macrophages and dendritic cells [24–26]. Importantly, the innate immune response enhances the development of adaptive (HLA-DQ2 or -DQ8 mediated) immunity to gluten [26]. A peptide from α2 gliadin (p31–43) has been described as an innate trigger for intestinal epithelial cells and especially in intestinal organ cultures [26], whereas other peptides have been found to stimulate cytokine production in rodent monocytes or macrophages [24, 25]. None of these peptides have been unanimously confirmed, no receptor for any of these peptides has been identified, and some believe that effects may be due to lipopolysaccharide contamination. Two recent studies that largely ruled out lipopolysaccharide implicate MyD88 and Toll like receptor 4 (TLR4) as receptors for innate responses to cereal proteins on monocytes, macrophages and dendritic cells [25, 27].

Intra-epithelial Lymphocytes and IL-15

Increased numbers (>25 per 100 epithelial cells) of intra-epithelial lymphocytes (IEL) are another characteristic though unspecific finding in c.d. The majority of IEL are cytotoxic CD8+ T cells that can express perforin/granzyme and FasL, but NK cells, γδ T cells and subsets of CD4+ cells represent other important populations [28]. Direct or indirect (e.g. via prior engagement of dendritic cells or macrophages) innate immune activation of intestinal epithelia by gluten induces their expression of the non-classical class I molecule MICA which acts as ligand for the heterodimeric NKG2D receptor on the above-mentioned IEL. This leads to activation of NKG2D and induction of the epithelial production of IL-15 [29]. Similarly,

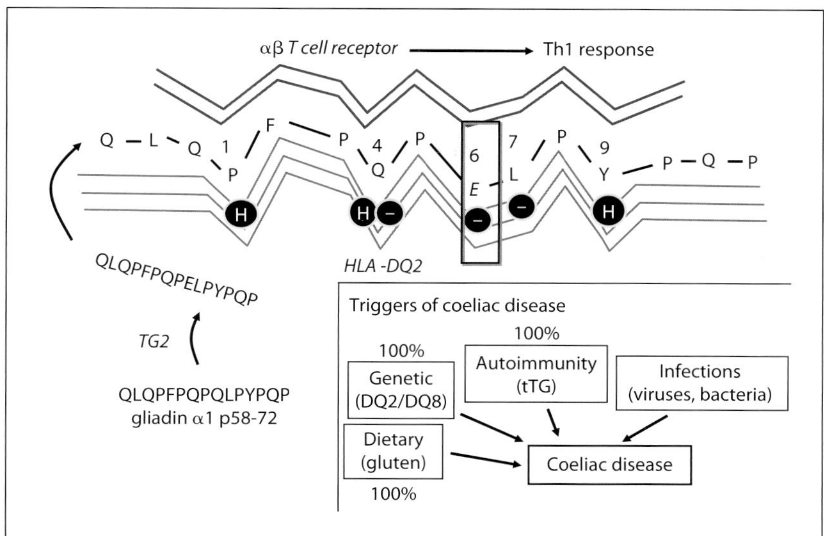

Fig. 2. Gluten, TG2 and antigen presentation. c.d. is driven by gluten in subjects carrying HLA-DQ2 (or -DQ8), and by the activity of the autoantigen TG2. Apart from other (minor) genetic contributions, mechanical and microbial triggers appear to contribute to disease manifestation. TG2 can deamidate gluten peptides, such as the immunodominant gliadin peptide α1 58-72 that harbours the core DQ2-binding sequence PFPQPQLPY (proline-phenylalanine-proline-glutamine-proline-glutamine-leucine-proline-tyrosine). This leads to the generation of a negatively charged residue of glutamate (E) in the peptide. Negatively charged residues in positions 4, 6 and 7, and hydrophic residues (H) in positions 1, 4 and 9 are preferred for optimal binding of the peptide in the antigen binding groove of DQ2. This results in stronger binding of the deamidated peptide to DQ2, and a more potent stimulation of gliadin specific T cells.

HLA-E becomes up-regulated on the intestinal epithelium, which activates a more restricted subset of IEL via the NKG2C receptor [30]. IEL can also assume an immunoregulatory capacity, since a subset of CD8+ TCRαβ+ IEL expresses the inhibitory NK receptor NKG2A, leading to secretion of the immunosuppressive TGF-β1 [31]. IL-15 which is both produced by activated intestinal epithelia and dendritic cells or macrophages is perhaps the most relevant activator of the innate and adaptive immune response in c.d. [28, 30–33]. It appears to act in concert with IL-21 which is produced by activated CD4+ Th1 T cells [34], and both cytokines should be central targets for the therapy of refractory c.d. type II (which is characterized by clonal growth of mainly IEL) and its overtly malignant variant, enteropathy associated T cell lymphoma (EATL).

Figures 2 and 3 summarize key concepts of c.d. pathogenesis. For an update on refractory c.d. and intestinal T cell lymphoma, regulatory T cell responses, preclinical and animal models of c.d. and non-invasive assessment of c.d. activity, refer to recent reviews and citations therein [4, 35–37].

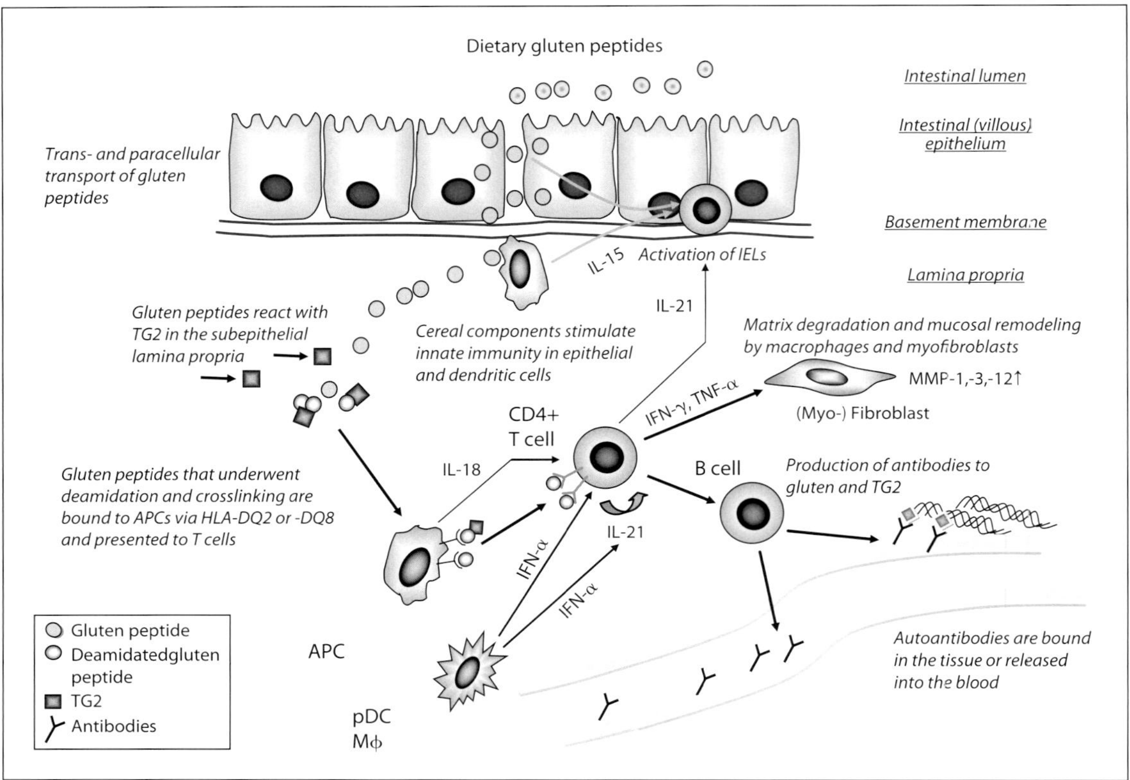

Fig. 3. Pathogenesis of c.d. Due to their high content of the amino acid proline, gluten peptides are resistant to gastric and intestinal proteases. Many of these peptides carry immune stimulatory peptides with a good fit for HLA-DQ2 or -DQ8. They reach the lamina propria either via epithelial transcytosis or an increased epithelial tight junction permeability. Here they encounter extracellular or cell-associated TG2 which is externalized by cells upon mechanical or inflammatory stress, leading to peptide crosslinking and deamidation. The deamidation creates potent immunostimulatory epitopes which are preferentially bound to HLA-DQ2 or -DQ8 on antigen presenting cells. Subsequently, gluten specific CD4+ T cells are activated, secreting mainly Th1 cytokines, such as IFN-γ, which induce the release and activation of matrix metalloproteinases by myofibroblasts. These metalloproteinases contribute to mucosal remodelling and villus atrophy. Th2 cytokines are produced during this process and induce the production of (auto-) antibodies to gluten and (haptemized) TG2. Other cytokines such as IL-18, IFN-α, and IL-21 support and maintain the Th1 polarization, and IL-15 links the innate with the adaptive immune system. This simplified sketch does not illustrate that T cells circulate to mesenteric lymph nodes where they encounter and are primed by the antigen presenting cells (APC, mainly dendritic cells), and from where they home back to the lamina propria. The homing is mediated by the lymphocyte chemokine receptor CCR9 and integrin α4β7, which are relevant targets for the therapy of refractory c.d. and EATL. IEL = Intra-epithelial lymphocyte; MΦ = macrophage; pDC = plasmacytoid dendritic cell. Modified from [4].

Genetic and Other Risk Factors

HLA-DQ2 or -DQ8 are necessary genetic predispositions for c.d. and are expressed in 30–35% of the populations where c.d. is prevalent, but only ~2–5% of gene carriers will develop c.d. This suggests there are other genetic and environmental factors that contribute to the manifestation of c.d. [28, 29]. Recently, genome-wide association studies have been performed in large cohorts of patients with c.d., their first degree relatives and matched controls. These studies demonstrated several additional risk genes and most of these are plausible, since they are related to T cell regulation and (intestinal) inflammation [38, 39]. The most prominent examples are the non-HLA genes CD28/CTLA-4/ICOS, IL2/IL21 (4q27), RGS1 (1q31), IL18RAP (2q11–2q12), CCR3 (3p21), IL12A (3q25–3q26), TAGAP (6q25) and SH2B3 (12q24).Yet the overall genetic risk contribution of these polymorphisms combined is estimated at only 3–4%, as compared to 30–35% for HLA-DQ2 or -DQ8. Moreover, early and massive exposure of infants to dietary gluten [40], infection with enteropathic viruses, or alterations of the bacterial flora [4, 41, 42] have been implicated as external risk factors for the development of clinically manifest c.d..

Oral Tolerance to Gluten

In analogy to inflammatory bowel disease, c.d. apparently results from up-regulation of a usually suppressed destructive T cell response to (normally harmless) luminal antigens (in this case, the defined gluten proteins). Thus studies aimed at restoring the regulatory T cell response to gluten hold promise as a causal treatment for c.d. and perhaps serve as template for the treatment of inflammatory bowel disease where the trigger is less well defined. Long-term restoration of oral tolerance to ingested gluten appears to be possible, since 13 of 61 young adults with childhood c.d. who remained on a gluten-free diet for several years did not develop c.d., despite having resumed a normal gluten containing diet in adolescence for an average of 10 years [43].

Intestinal Gluten Transport

It is largely unclear how the luminal immunogenic gluten peptides that survive intestinal digestion reach the lamina propria. They may either traverse the intestinal epithelial lining through defective tight junctions (paracellular pathway) [44], or be actively transported (epithelial transcytosis). The latter has been shown to be increased in the inflamed mucosa of patients with c.d. [44–48]. A possible mechanism is that gluten peptides reach the basal layer of the epithelia in association with luminal anti-gluten IgA via retrotranscytosis from the apical side [48], but other mechanisms could rather be implicated. Thus, there may exist a yet unidentified gluten transporter in the epithelium, or peptides may be sampled by interdigitating lamina propria dendritic cells. The in vivo existence of interdigitating dendritic cells has been shown in mice but not yet in humans [49]. Furthermore, sampling of gluten peptides by antigen-presenting cells may occur via the specialized microfold cells that are part of the follicle-associated epithelium of the mucosal-associated lymphoid tissue [50]. Identification of the major mechanism will open novel possibilities for pharmacological inhibition of gluten uptake.

Novel Therapies for Coeliac Disease

Most patients with c.d. can be treated effectively with a strict gluten-free diet. However, this diet is difficult to maintain, often has a lower nutritional value, is costly, can lead to social restrictions, and is not easily available in many countries. Therefore, given the broad usage of wheat and related cereal products in most countries, and hidden undeclared gluten as an additive in many foods, compliance with the gluten-free diet can be low and the danger of at least low-level exposure in strictly adherent patients is high. Importantly, a sizable proportion of patients are highly sensitive to trace amounts of gluten, which necessitates non-dietary, adjunctive treatments [2–4, 51]. Furthermore, such treatments are necessary for refractory c.d. type I (which can also be treated by common immunosuppressants), and especially refractory c.d. type II (which does not respond well to immunosuppression), and EATL (which is fatal in most of the patients, despite chemotherapy or bone marrow transplantation). Novel therapies should be of low risk and reasonable cost in patients with milder forms of the disease, and must be highly effective in patients with severe disease, with an accepted higher risk and cost. Table 1 summarizes therapies that are either already in phase I or II clinical studies, in a preclinical stage (with proven efficacy in T cell assays, organ cultures or rodent models of c.d.), or being discussed for evaluation. The therapies that can be classified according to their site of action and subdivided as to their molecular mechanisms and targets are discussed in detail in [4].

Conclusions

Taken together, the research into c.d. has made tremendous progress in the last two decades. Importantly, c.d. serves as a model gastrointestinal disease of immune dysregulation, since its triggers, its genetic and environmental predispositions, and its innate, adaptive and autoimmune components are relatively well understood. This makes the preclinical and final clinical validation of targeted therapies feasible, and indeed such therapies are expected to make it into the clinic in the next few years. Therefore, advances in c.d. will also spur translational research in other immune-mediated disorders of the gastrointestinal tract and of other organs.

Table 1. Novel therapies for c.d.

Targeted pathology	Drug/modification	State of development
Intraluminal therapies		
Wheat variants	Wheat with lower immunogenicity: (a) Ancestral wheat variants (b) Genetically modified wheat	Preclinical, tested on biopsies and gliadin-reactive T cell lines
Flour/dough	Pretreatment with lactobacilli leading to (partial) digestion of immunodominant gluten peptides	Clinical trial with 17 patients
	Transamidation and thus inactivation of immunodominant gliadin sequences	Preclinical, tested on gliadin reactive T cell lines
Ingested immunodominant gliadin peptides	Prolyl endopeptidases that degrade immunodominant gluten peptides, derived from:	
	(a) *Aspergillus niger*	Phase I clinical trial (NCT00810654)
	(b) *Sphingomonas capsulate* in combination with endopeptidase-B2 from germinating barley	Phase I clinical trial (NCT00626184)
	Intraluminal gliadin binding by a polymethacrylate-polystyrene-sulfonic acid polymer	Preclinical
	Gluten neutralizing cow's milk antibodies that resist gastric and intestinal degradation	Preclinical
Transepithelial uptake		
Epithelial tight junctions	Intestinal epithelial cell cholera toxin (ZOT) receptor antagonist AT1001	Phase IIb clinical trial (NCT00889473)
Dampening of the adaptive immune response to gluten		
Tissue transglutaminase	Transglutaminase inhibitors that prevent gluten deamidation	Preclinical, tested ex vivo on biopsies
Regulatory T cells?	'Inhibitory' innate gluten peptides	Preclinical, tested on biopsies and gliadin reactive T cell lines
HLA-DQ2	Blocking DQ2-analogues	Preclinical, tested on gliadin reactive T cell lines
Induction of an intestinal Th2 response	Hookworm infection	Phase II clinical trial (NCT00671138)

Table 1. Continued

Targeted pathology	Drug/modification	State of development
Desensitization to 3 major immunogenic epitopes	Gluten 'vaccination' (Nexvax2)	Phase I–II clinical trial (NCT00879749)
Biologicals (systemic T cell or cytokine blockers)		
Small intestine homing T cells	CCR9 antagonists (CCX282-B, CCX025)	Phase II clinical trial planned (NCT00540657)
Gut homing T cells	Anti-integrin α4β7 (LDP-02)	Phase II clinical trial for Crohn's disease (NCT00655135)
Clonal IEL	(a) Anti-IL-15 (AMG 714)	Phase II clinical trial for rheumatoid arthritis (NCT00433875)
	(b) Anti-Jak3 (CP-690-550)	Phase II clinical trial for rheumatoid arthritis and transplant rejection (NCT00550446, NCT00658359)
Clonal intestinal T cells	(a) Autologous bone marrow transplantation	Clinical trial on patients with EATL
	(b) Mesenchymal stem cell transplantation (Prochymal)	Phase II clinical trial for Crohn's disease (NCT00294112)
Mucosal destruction in refractory celiac disease	(a) Anti-TNF-α, anti-IFNγ (HuZAF)	Case reports in celiac disease Phase II clinical trial for Crohn's disease (NCT00072943)
	(b) Anti-CD52 (Alemtuzumab)	Case reports in c.d.

Some of the therapies above are in phase I or II clinical trials for c.d., others are currently being tested for other indications. Most approaches have only reached the preclinical stage, either with proof of concept testing on gluten-reactive T cell lines or biopsies from patients with c.d., or in (as yet imperfect) animal models.
Modified from [4], which contains the relevant references and a more in-depth discussion of the therapies.

Acknowledgments

The author's work on coeliac disease is supported by grant 1R21DK073254-02 from the National Institutes of Health.

References

1 Green PH, Cellier C: Celiac disease. N Engl J Med 2007;357:1731–1743.

2 Sollid LM, Lundin KE: Diagnosis and treatment of celiac disease. Mucosal Immunol 2009;2:3–7.

3 Di Sabatino A, Corazza GR: Coeliac disease. Lancet 2009;373:1480–1493.

4 Schuppan D, Junker Y, Barisani D: Celiac Disease: From Pathogenesis to novel therapies. Gastroenterology, in press.

5 Catassi C, Ratsch IM, Fabiani E, et al: Coeliac disease in the year 2000: exploring the iceberg. Lancet 1994;343:200–203.

6 Fasano A, Berti I, Gerarduzzi T, et al: Prevalence of celiac disease in at-risk and not-at-risk groups in the United States: a large multicenter study. Arch Intern Med 2003;163:286–292.

7 Vilppula A, Kaukinen K, Luostarinen L, et al: Increasing prevalence and high incidence of celiac disease in elderly people: a population-based study. BMC Gastroenterol 2009;9:49.

8 Ventura A, Magazzu G, Greco L: Duration of exposure to gluten and risk for autoimmune disorders in patients with celiac disease: SIGEP Study Group for Autoimmune Disorders in Celiac Disease. Gastroenterology 1999;117:297–303.

9 Al-Toma A, Verbeek WH, Hadithi M, et al: Survival in refractory coeliac disease and enteropathy-associated T-cell lymphoma: retrospective evaluation of single-centre experience. Gut 2007;56:1373–1378.

10 Malamut G, Afchain P, Verkarre V, et al: Presentation and long-term follow-up of refractory celiac disease: comparison of type I with type II. Gastroenterology 2009;136:81–90.

11 Wieser H: Chemistry of gluten proteins. Food Microbiol 2007;24:115–119.

12 Pender SL, Tickle SP, Docherty AJ, et al: A major role for matrix metalloproteinases in T cell injury in the gut. J Immunol 1997;158:1582–1590.

13 Dieterich W, Ehnis T, Bauer M, et al: Identification of tissue transglutaminase as the autoantigen of celiac disease. Nat Med 1997;3:797–801.

14 Elli L, Bergamini CM, Bardella MT, et al: Transglutaminases in inflammation and fibrosis of the gastrointestinal tract and the liver. Dig Liver Dis 2009;41:541–550.

15 Schuppan D, Dieterich W, Ehnis T, et al: Identification of the autoantigen of celiac disease. Ann NY Acad Sci 1998;859:121–126.

16 Fleckenstein B, Molberg O, Qiao SW, et al: Gliadin T cell epitope selection by tissue transglutaminase in celiac disease: role of enzyme specificity and pH influence on the transamidation versus deamidation process. J Biol Chem 2002;277:34109–34116.

17 Molberg O, McAdam SN, Korner R, et al: Tissue transglutaminase selectively modifies gliadin peptides that are recognized by gut-derived T cells in celiac disease. Nat Med 1998;4:713–717.

18 van de Wal Y, Kooy Y, van Veelen P, et al: Selective deamidation by tissue transglutaminase strongly enhances gliadin-specific T cell reactivity. J Immunol 1998;161:1585–1588.

19 Anderson RP, Degano P, Godkin AJ, et al: In vivo antigen challenge in celiac disease identifies a single transglutaminase-modified peptide as the dominant A-gliadin T-cell epitope. Nat Med 2000;6:337–342.

20 Arentz-Hansen H, McAdam SN, Molberg O, et al: Celiac lesion T cells recognize epitopes that cluster in regions of gliadins rich in proline residues. Gastroenterology 2002;123:803–809.

21 Shan L, Molberg O, Parrot I, et al: Structural basis for gluten intolerance in celiac sprue. Science 2002; 297:2275–2279.

22 Vader LW, Stepniak DT, Bunnik EM, et al: Characterization of cereal toxicity for celiac disease patients based on protein homology in grains. Gastroenterology 2003;125:1105–1113.

23 Qiao SW BE, Molberg O, Xia J, Fleckenstein B, Khosla C, Sollid LM: Antigen presentation to celiac lesion-derived T cells of a 33-mer gliadin peptide naturally formed by gastrointestinal digestion. J Immunol 2004;173:1757–1762.

24 Cinova J, Palova-Jelinkova L, Smythies LE, et al: Gliadin peptides activate blood monocytes from patients with celiac disease. J Clin Immunol 2007; 27:201–209.

25 Thomas KE, Sapone A, Fasano A, et al: Gliadin stimulation of murine macrophage inflammatory gene expression and intestinal permeability are MyD88-dependent: role of the innate immune response in Celiac disease. J Immunol 2006;176: 2512–2521.

26 Londei M, Ciacci C, Ricciardelli I, et al: Gliadin as a stimulator of innate responses in celiac disease. Mol Immunol 2005;42:913–918.

27 Junker Y, Leffler DA, Wieser H, et al: Gliadin activates monocytes, macrophages and dendritic cells in vitro and in vivo via Toll like receptor 4. Gastroenterology 2009;36(suppl 1):M2022.

28 Di Sabatino A, Ciccocioppo R, Cupelli F, et al: Epithelium derived interleukin 15 regulates intraepithelial lymphocyte Th1 cytokine production, cytotoxicity, and survival in coeliac disease. Gut 2006;55:469–477.

29 Hue S, Mention JJ, Monteiro RC, et al: A direct role for NKG2D/MICA interaction in villous atrophy during celiac disease. Immunity 2004;21:367–377.

30 Terrazzano G, Sica M, Gianfrani C, et al: Gliadin regulates the NK-dendritic cell cross-talk by HLA-E surface stabilization. J Immunol 2007;179:372–381.

31 Bhagat G, Naiyer AJ, Shah JG, et al: Small intestinal CD8+ TCRgammadelta+ NKG2A+ intraepithelial lymphocytes have attributes of regulatory cells in patients with celiac disease. J Clin Invest 2008;118: 281–293.

32 Mention JJ, Ben Ahmed M, Begue B, et al: Interleukin 15: a key to disrupted intraepithelial lymphocyte homeostasis and lymphomagenesis in celiac disease. Gastroenterology 2003;125:730–745.

33 Maiuri L, Ciacci C, Vacca L, et al: IL-15 drives the specific migration of CD94+ and TCR-gammadelta+ intraepithelial lymphocytes in organ cultures of treated celiac patients. Am J Gastroenterol 2001; 96:150–156.
34 Meresse B, Verdier J, Cerf-Bensussan N: The cytokine interleukin 21: a new player in coeliac disease? Gut 2008;57:879–881.
35 Marietta E, Schuppan D, Murray JA: In vitro and in vivo models of celiac disease. Exp Opin Drug Discov 2009, in press.
36 Meresse B, Cerf-Bensussan N: Innate T cell responses in human gut. Semin Immunol 2009;21: 121–129.
37 Jabri B, Sollid LM: Mechanisms of disease: immunopathogenesis of celiac disease. Nat Clin Pract Gastroenterol Hepatol 2006;3:516–525.
38 Hunt KA, Zhernakova A, Turner G, et al: Newly identified genetic risk variants for celiac disease related to the immune response. Nat Genet 2008; 40:395–402.
39 Dubois PC, van Heel DA: Translational mini-review series on the immunogenetics of gut disease: immunogenetics of coeliac disease. Clin Exp Immunol 2008;153:162–173.
40 Hunt KA, Zhernakova A, Turner G, et al: Newly identified genetic risk variants for celiac disease related to the immune response. Nat Genet 2008; 40:395–402.
41 Pavone P, Nicolini E, Taibi R, et al: Rotavirus and celiac disease. Am J Gastroenterol 2007;102:1831.
42 Collado MC, Donat E, Ribes-Koninckx C, et al: Specific duodenal and faecal bacterial groups associated with paediatric coeliac disease. J Clin Pathol 2009;62:264–269.
43 Matysiak-Budnik T, Malamut G, de Serre NP, et al: Long-term follow-up of 61 coeliac patients diagnosed in childhood: evolution toward latency is possible on a normal diet. Gut 2007;56:1379–1386.
44 Clemente MG, De Virgiliis S, Kang JS, et al: Early effects of gliadin on enterocyte intracellular signalling involved in intestinal barrier function. Gut 2003;52:218–223.
45 Schumann M, Richter JF, Wedell I, et al: Mechanisms of epithelial translocation of the alpha(2)-gliadin-33mer in coeliac sprue. Gut 2008;57:747–754.
46 Zimmer KP, Poremba C, Weber P, et al: Translocation of gliadin into HLA-DR antigen containing lysosomes in coeliac disease enterocytes. Gut 1995;36: 703–709.
47 Matysiak-Budnik T, Candalh C, Dugave C, et al: Alterations of the intestinal transport and processing of gliadin peptides in celiac disease. Gastroenterology 2003;125:696–707.
48 Matysiak-Budnik T, Moura IC, Arcos-Fajardo M, et al: Secretory IgA mediates retrotranscytosis of intact gliadin peptides via the transferrin receptor in celiac disease. J Exp Med 2008;205:143–54.
49 Niess JH, Brand S, Gu X, et al: CX3CR1-mediated dendritic cell access to the intestinal lumen and bacterial clearance. Science 2005;307:254–258.
50 Man AL, Prieto-Garcia ME, Nicoletti C: Improving M cell mediated transport across mucosal barriers: do certain bacteria hold the keys? Immunology 2004;113:15–22.
51 Sollid LM, Khosla C: Future therapeutic options for celiac disease. Nat Clin Pract Gastroenterol Hepatol 2005;2:140–147.

Detlef Schuppan, MD, PhD
Division of Gastroenterology and Hepatology
Beth Israel Deaconess Medical Center, Harvard Medical School
330 Brookline Ave, Boston, MA 02215 (USA)
Tel. +1 617 667 2371, Fax +1 617 667 2767, E-Mail dschuppa@bidmc.harvard.edu

Mayerle J, Tilg H (eds): Clinical Update on Inflammatory Disorders of the Gastrointestinal Tract.
Front Gastrointest Res. Basel, Karger, 2010, vol 26, pp 95–107

Anti-TNF Therapy in Inflammatory Bowel Diseases

Gionata Fiorino[a,b] · Silvio Danese[b] · Laurent Peyrin-Biroulet[c]

[a]GI Unit, Dipartimento di Scienze Cliniche, Policlinico Umberto I, Rome, [b]Istituto Clinico Humanitas, IRCCS in Gastroenterology, Milan, Italy; [c]Inserm, U954 and Department of Hepato-Gastroenterology, University Hospital of Nancy, Vandoeuvre-lès-Nancy, France

Abstract

The introduction in the mid-1990s of anti- TNF-α agents changed the treatment of inflammatory bowel diseases (IBD), Crohn's disease and ulcerative colitis refractory to conventional medications (corticosteroids, immunomodulators). This review summarizes current data on efficacy, safety and indications of anti-TNF therapy in IBD, and discusses future directions for these agents in IBD. We searched Medline, the Cochrane Library and Embase for relevant studies. Infliximab, adalimumab and certolizumab are more effective than placebo for induction and maintenance of remission in luminal Crohn's disease. Infliximab is effective for maintenance of fistula closure in Crohn's disease, and adalimumab is also likely to have this activity. Only infliximab is approved by the US Food and Drug Administration for ulcerative colitis. Only adalimumab has demonstrated its efficacy to induce remission after infliximab failure in Crohn's disease in a randomized controlled trial. Anti-TNF therapy leads to mucosal healing, reduces hospitalizations and surgeries, and improves patients' quality of life. Safety data indicate that serious infections occur in 2–4% of patients treated with anti-TNF therapy, with no statistical difference when compared to controls. The risk of rare events, such as malignancies and lymphoma, in IBD patients treated with anti-TNF agents, requires a longer duration of follow-up. It is concluded that currently, the risk-benefit ratio of anti-TNF therapy supports its use in IBD. Several questions remain to be answered: are TNF antagonists truly disease-modifying agents, should mucosal healing be used in clinical practice, and should anti-TNF therapy be used alone or in combination with immunomodulators in the long term?

Inflammatory bowel disease (IBD) encompasses Crohn's disease (CD) and ulcerative colitis (UC). The aetiology and pathomechanisms leading to the development of these complex disorders remain poorly understood. Both innate and adaptive immunity seems to be implicated in triggering and maintaining pathological intestinal inflammation [1]. The pro-inflammatory cytokine TNF-α may have a key role in the pathogenesis of these chronic conditions. Mice deficient in TNF-α develop Crohn's-like lesions [1]. Higher levels of TNF have also been found within intestinal mucosa of IBD patients compared to healthy subjects [2].

Since the mid-1990s, the role of TNF-α as a major mediator of inflammation in the human gut has been supported by the efficacy of a new class of molecules directed against TNF in IBD, namely the anti-TNF agents. The first patients were treated with infliximab in the mid-1990s [3]. Six anti-TNF drugs have been studied in luminal CD: infliximab, adalimumab, certolizumab pegol, CDP571, etanercept and onercept [4]. CDP571, onercept and etanercept are not effective in CD and are not used in the treatment of IBD [4].

In 1998, infliximab became the first anti-TNF agent to be approved for luminal CD and, later, for fistulizing CD and UC. This monoclonal chimeric antibody, administered intravenously, has dramatically improved the treatment of IBD that is refractory to conventional medications (corticosteroids and immunomodulators). The murine part of the molecule may lead to immunogenicity, possibly resulting in loss of response, even though the impact of anti-infliximab antibodies on efficacy and safety remains uncertain [5]. Optimization of therapy by reducing the interval between injections and/or increasing dose should be considered in cases of loss of response or intolerance to infliximab before switching to another anti-TNF agent.

More recently, 2 other anti-TNF agents, both administered subcutaneously, have been approved by the US Food and Drug Administration (FDA) for the treatment of luminal CD. They are certolizumab pegol, a humanized anti-TNF pegylated Fab′ fragment, and adalimumab, a fully human monoclonal antibody [4]. Importantly, we should keep in mind that all anti-TNF agents have the potential for immunogenicity. In Europe, certolizumab has not yet received approval for IBD. The molecular features and the main characteristics of the 3 FDA-approved anti-TNF agents for luminal CD are shown in figures 1 and 2, respectively.

The aim of this review is to summarize the available data on efficacy, safety and indications of anti-TNF therapy in IBD, and to discuss future directions regarding the use of these biological agents for lifelong, incurable diseases.

Methods

A review of the literature was conducted searching for the terms 'tumour necrosis factor', 'anti-TNF', 'infliximab', 'adalimumab', 'etanercept', 'certolizumab', 'onercept', 'etanercept', 'CDP 571', and 'biological agents' matched with 'Crohn's disease', 'ulcerative colitis', 'inflammatory bowel disease', 'intolerance', 'loss of response', 'side effects' and 'safety' in the Embase, Pubmed and Cochrane databases, as previously described [4]. Relevant articles in English and other languages were reviewed.

Results

Efficacy as a First-Line Anti-TNF Therapy

The main results of the reviewed trials are summarized in table 1.

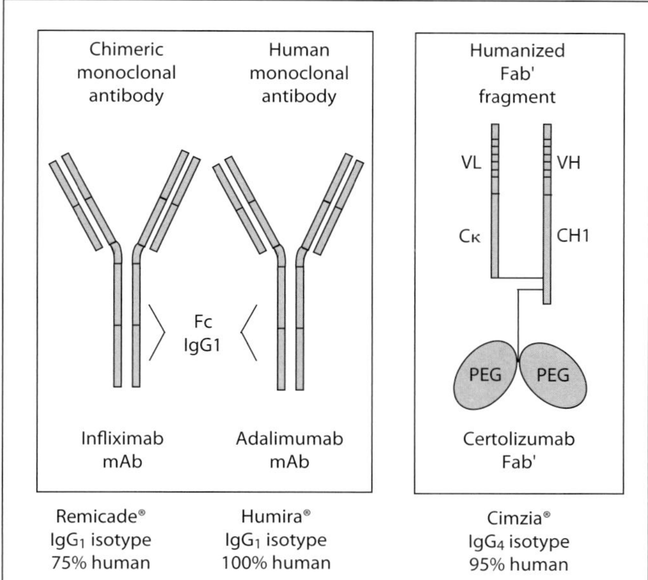

Fig. 1. Schematic representation and molecular features of the 3 anti-TNF agents approved by the FDA for the treatment of luminal Crohn's disease: infliximab, adalimumab, certolizumab pegol.

	Modes of action			Mode of administration	Half-life (days)	Interval between injections (weeks)
	TNF neutralization	Apoptosis	Others			
Infliximab	+	+	CD40/ CD40L, ADCC, CDC	i.v.	8–9.5	8
Adalimumab	+	+	ADCC, CDC	s.c.	12–14	2
Certolizumab	+	No (?)	No	s.c.	14	4

Fig. 2. Main characteristics of the 3 anti-TNF agents approved by the FDA for the treatment of luminal Crohn's disease: infliximab, adalimumab, certolizumab pegol. CD40/CD40L = Down-regulation of the CD40/CD40L pathway that mediates innate immunity and inflammatory processes; ADCC = antibody-dependent cell-mediated cytotoxicity; CDC = complement-dependent cytotoxicity.

Luminal Crohn's Disease

Since 1997, fourteen randomized placebo-controlled trials evaluating the efficacy of anti-TNF therapy in luminal CD have been published as full-length papers [for review see 4]. These trials included a total of 3,955 patients: 308 were treated with infliximab, 713 with adalimumab and 830 with certolizumab [4].

Table 1. Summary of efficacy as a first-line anti-TNF therapy and marketing approval for anti-TNF agents in IBD

	Efficacy			FDA approval		
	luminal CD	fistulizing CD	UC	luminal CD	fistulizing CD	UC
Infliximab	yes	yes	yes	yes	yes	yes
Adalimumab	yes	yes[a]	?	yes	no	no
Certolizumab	yes	?	?	yes	no	no

[a] Efficacy of adalimumab in fistulizing CD needs to be confirmed in randomized controlled trials evaluating fistula closure as a primary endpoint.

Induction

Infliximab was effective in inducing clinical remission (defined by a Crohn's Disease Activity Index <150) in 33% of patients compared with only 4% of the placebo group at week 4 ($p = 0.005$) [6]. Adalimumab was also significantly more effective than placebo to induce clinical remission (36 vs. 12%, respectively, $p < 0.001$) at doses of 160 mg at week 0 followed by 80 mg at week 2 [7]. Of note, an adalimumab induction dosing regimen with 80 mg at week 0 followed by 40 mg at week 2 was not more effective than placebo to induce remission at week 4 [7]. In a randomized trial enrolling 292 patients with moderate to severe CD, certolizumab pegol 400 mg at weeks 0, 4 and 8 was not more effective than placebo to induce clinical response at week 12 [8]. More recently, in a large randomized placebo-controlled trial ($n = 662$), response rates at week 6 were 35% in the certolizumab group (injections at weeks 0, 2 and 4 and then every 4 weeks) and 27% in the placebo group ($p = 0.02$) [9]. However, at week 6, the rates of remission in the 2 groups did not differ significantly [9].

Maintenance

Efficacy of infliximab, adalimumab and certolizumab to maintain clinical remission was evaluated in 5 randomized placebo-controlled trials [for review see 4].

A first randomized, placebo-controlled trial established efficacy of retreatment with 10 mg/kg infliximab every 8 weeks in 73 CD patients [10]. A second large randomized placebo-controlled trial enrolled 573 patients with CD who responded to a single 5 mg/kg infliximab infusion at week 0 and who received infliximab infusions (5 or 10 mg/kg) or placebo at weeks 2 and 6 and then every 8 weeks until week 46 [11]. At week 30, 21% of patients were in clinical remission in the placebo group, compared to 39–45% of patients in the infliximab groups (all comparisons were statistically significant). The median time to loss of response was also significantly lower in the placebo group than in the active arms (19 vs. 38 weeks to >54 weeks, respectively). No differences in clinical efficacy were observed between the 5 mg/

kg and 10 mg/kg infliximab groups [11]. Hence, the FDA-approved dose is 5 mg/kg for infliximab induction (at weeks 0, 2 and 6) and maintenance therapy (every 8 weeks). Comparison of scheduled and episodic treatment strategies of infliximab in CD clearly demonstrated that compared to episodic therapy, scheduled infliximab therapy is associated with higher rates of clinical response and mucosal healing, better quality of life, fewer hospitalizations and decreased immunogenicity [12–14]. Scheduled infliximab therapy is thus recommended in IBD patients responding to induction therapy.

Adalimumab, 40 mg every other week, was also significantly more effective than placebo in maintaining clinical remission at week 56 (36 vs. 16%, respectively) in initial responders to open-label induction therapy with adalimumab 80 mg at week 0 followed by 40 mg at week 2 [15]. In this trial, there were no significant differences in efficacy between adalimumab treatment every other week versus weekly dosing. Similar to infliximab, scheduled treatment with adalimumab was more effective than episodic treatment [16].

Infliximab and adalimumab have also been shown to be more effective than placebo in maintaining steroid-free remission at 1 year [11, 15].

Regarding certolizumab, 2 randomized placebo-controlled trials were conducted to evaluate its efficacy at 6 months [9, 17]. In the PRECISE 1 study, in which patients were randomly assigned to receive either certolizumab or placebo, the rates of remission in the 2 groups did not differ significantly at 6 months [9]. In the PRECISE 2 study, among patients responding to open-label induction therapy, clinical remission rates at week 26 were significantly higher in the certolizumab group than in the placebo group (48% vs. 29%, respectively) [17].

Fistulizing Crohn's Disease

The advent of anti-TNF therapy has revolutionized the treatment of fistulising CD. Only infliximab and adalimumab have shown efficacy in fistulising CD in randomized controlled trials [15, 18–20]. However, fistula closure was a secondary objective in the adalimumab maintenance trial [15, 20]. Therefore, infliximab remains the only anti-TNF approved by the FDA for the treatment of fistulising CD.

Induction

A landmark study included 94 CD patients who had draining abdominal or perianal fistulas [19]. Patients were randomly assigned to receive placebo, 5 mg/kg of infliximab, or 10 mg/kg of infliximab at weeks 0, 2 and 6. Sixty-eight percent of the patients who received 5 mg/kg infliximab achieved the primary end point (reduction of 50 percent or more from baseline in the number of draining fistulas observed at 2 or more consecutive study visits), as compared with only 26% in the placebo group ($p = 0.002$). In addition, 55% of the patients assigned to receive 5 mg/kg infliximab had complete fistula closure, as compared with only 13% of the patients assigned to placebo ($p = 0.001$) [19].

The efficacy of adalimumab and certolizumab to induce fistula closure in CD has not been specifically studied.

Maintenance

In the ACCENT II study enrolling 306 CD patients with one or more draining abdominal or perianal fistulas, 36% of patients who had responded to induction therapy and received maintenance infliximab therapy maintained complete fistula closure at week 54 compared with 19% of patients receiving placebo ($p = 0.009$) [18].

In a secondary analysis of the CHARM trial, complete fistula closure was achieved in a greater percentage of adalimumab-treated patients ($n = 70$) versus those receiving placebo ($n = 47$) at week 56 (33 and 13% for combined adalimumab groups and placebo group, respectively, $p = 0.016$) [15]. In an open-label extension of the CHARM trial that enrolled the 31 patients with healed fistulas at week 56, 90% maintained complete fistula closure at the end of an additional year of follow-up [21].

Large randomized controlled trials evaluating fistula closure, as a primary endpoint, are awaited for adalimumab and certolizumab.

Ulcerative Colitis

The ACT 1 and 2 trials, which were large randomized, placebo-controlled studies, evaluated the efficacy of infliximab for induction and maintenance therapy in UC [21]. In each study, 364 patients with moderate-to-severe UC received placebo or infliximab (5 mg or 10 mg/kg) at weeks 0, 2, and 6 and then every 8 weeks through week 46 (in ACT 1) or week 22 (in ACT 2). In ACT 1, at week 8, 69% of patients in the group receiving 5 mg of infliximab had had a clinical response, as compared with 37% of patients in the placebo group ($p < 0.001$). In ACT 2, at week 8, 64.5% of patients in the group receiving 5 mg of infliximab had had a clinical response, as compared with 29% of patients in the placebo group ($p < 0.001$). The proportions of patients who were in clinical remission and discontinued corticosteroids at 6 months or 1 year were higher in the infliximab groups than in the placebo groups [21].

The efficacy of adalimumab and certolizumab as a first-line anti-TNF therapy for UC is unknown.

Efficacy of a Second Anti-TNF Agent after Infliximab Failure

Adalimumab in Luminal Crohn's Disease

A 4-week, randomized, placebo-controlled trial enrolled 325 adults with moderate-to-severe CD who had loss of response or intolerance to infliximab. At week 4, 21% of patients in the adalimumab group achieved clinical remission versus only 7% in the placebo group ($p < 0.001$) [22].

A 3-year experience from a referral centre in France showed that the probability of maintaining clinical response under adalimumab as a second-line therapy was 51% at 130 weeks. Eight of 10 patients were able to taper steroid therapy [23].

The efficacy of adalimumab as a second-line anti-TNF therapy to maintain clinical remission in patients with luminal CD needs to be confirmed in large randomized controlled trials.

Adalimumab in Ulcerative Colitis

Three uncontrolled studies evaluated the efficacy of adalimumab in UC patients who failed infliximab therapy [24, 25]. An induction trial of over 10 patients showed a small rate of induction of remission: 1 patient achieved remission, 3 had clinical response, 6 did not respond at week 4 [19]. In another uncontrolled trial from the same group (University Hospital of Nancy, France), after a median duration of follow-up of 42 weeks, the probability of maintaining adalimumab was 32.5% at 23 months [25]. Six of 13 patients (46.2%) underwent colectomy during the study [25]. More recently, in an open-label study of adalimumab in 20 UC patients including those with prior loss of response or intolerance to infliximab (n = 13), the rates of clinical remission were 5% and 20% at weeks 8 and 24, respectively [26].

Certolizumab in Luminal Crohn's Disease

In a large open-label study enrolling 539 patients who had no response, lost response, or intolerance to infliximab, 62% of patients achieved clinical response and 39% achieved clinical remission at week 6 with certolizumab [27]. These results are comparable to those in the PRECISE 2 study [17], suggesting that prior infliximab exposure may not reduce response to certolizumab for induction of remission.

Overall, the efficacy of adalimumab and certolizumab after infliximab failure needs to be confirmed in randomized controlled trials.

Safety

Data from Referral Centres

In a Mayo Clinic study, after a median follow-up of 17 months, 6% of 500 patients experienced a serious adverse event judged to be related to infliximab therapy [28]. Acute infusion reactions occurred in 3.8% of patients, serum sickness-like disease attributed to infliximab in 2.8% of patients. In addition, 8.2% had an infectious event attributed to infliximab, 4% had a serious infection, and 1 patient developed a new demyelinating disorder [28].

More recently, in Leuven, Belgium, after a median follow-up of 58 moths, severe adverse events occurred in 13% of 734 IBD patients treated with infliximab; which was similar to the adverse event rate in controls (p = 0.45) [29]. There was no difference between IBD and control groups in mortality, malignancies and infection rate. The most commonly reported systemic side effects were skin eruptions with psoriasiform eruptions (seen in 20% of patients). This increased risk of psoriasiform lesions represents a growing concern in clinical practice. Concomitant steroids use was the only independent risk factor for infectious complications [29].

The TREAT registry (Crohn's Therapy, Resource, Evaluation and Assessment Tool), which includes 3,179 CD patients who received infliximab, also demonstrated that this agent was not an independent predictor of serious infections [30]. Factors independently associated with serious infections included prednisone use, narcotic analgesic use, and moderate-to-severe disease activity. Mortality rates were similar between patients who received infliximab and those treated with other drugs [30]. Immunosuppressive medications, when used in combination with anti-TNF agents, may be associated with increased risk of opportunistic infections in IBD patients [31].

Data from Randomized Controlled Trials

In a meta-analysis of 21 placebo-controlled trials enrolling 5,356 individuals, anti-TNF therapy did not increase the risk of death, malignancy or serious infection when compared to control arms. Notably, there was no difference in the frequency of serious infections between anti-TNF and control groups (2.09 vs. 2.13%, respectively) [4]. However, a longer duration of follow-up and a larger number of patients are required to better assess the safety profile of anti-TNF agents in CD.

Colombel et al. [32] performed a pooled analysis of 6 global clinical trials evaluating induction and maintenance of clinical remission with adalimumab in 3,160 patients with moderately to severely active Crohn's disease. The types and frequency of adverse events, serious adverse events and adverse events of interest (infections, serious infectious, malignancies, demyelinating disorders, lupus-like syndrome, deaths) were similar to placebo [32], and confirmed the data from the meta-analysis by Peyrin-Biroulet et al. [4].

Recently, Siegel et al. [33] evaluated the risk of lymphoma in CD patients treated with both immunomodulators and TNF antagonists by performing a meta-analysis of all types of studies (randomized controlled trials, prospective or retrospective cohort studies, case series of consecutive patients, published articles or meeting abstracts) [33]. There was a significant increased risk of lymphoma in patients who had received anti-TNF agents (standardized incidence ratio 3.23; 95% CI 1.5–6.9) when compared to a population-based registry of CD[33]. These results should be interpreted with caution due to methodological limitations [34]. Notably, 10 out of 13 patients who developed lymphoma were also on immunomodulator therapy. In addition, only studies with a minimum follow-up of 48 weeks were included. Lymphoma may occur during the first year of treatment. Overall, these 2 meta-analyses do not allow definitive conclusions to be drawn regarding the risk of lymphoma in CD patients receiving anti-TNF therapy.

In contrast to the article by Toruner et al. [31], subgroup analyses across 4 randomized placebo-controlled trials including a total of 1,383 patients from ACCENT I and II (luminal and fistulizing CD trials) and ACT 1 and 2 (UC trials), in which patients were treated with infliximab maintenance therapy, showed that infection and serious infection rates were generally similar in patients who received maintenance therapy with or without concomitant immunomodulators [35].

Collectively, these data indicate the good overall safety of anti-TNF therapy and support their use in IBD.

Novel Therapeutic Strategies and Goals

'Step-Up' or 'Top-Down' Approach?

In IBD, the goal is no longer simply symptom control, but rather alteration of the natural history of disease. Whether anti-TNF agents are disease modifying agents remains unknown. A trial evaluated efficacy of early and aggressive treatment with infliximab in CD, the so-called 'top-down' approach. This 2-year open-label randomized trial enrolled 133 patients with early CD [36]. The 'top-down group' was assigned to early combined immunosuppression and received infusions of infliximab at weeks 0, 2 and 6, in combination with azathioprine. The 'step-up' group received conventional treatment with either methylprednisolone or budesonide; azathioprine and thereafter infliximab therapy could be introduced in case of therapy failure. At 52 weeks, 61.5% in the early combined immunosuppression group were in remission compared with 42.2% in the conventional management group ($p = 0.0278$). However, after week 52, the proportion of patients in remission did not differ between the 2 groups [36]. Even though scheduled maintenance treatment with infliximab is superior to the episodic treatment that was used in this trial, the indiscriminate use of infliximab as a first-line therapy may result in the over-treatment of patients with CD. Only certain subgroups of patients, such as those less than 40 years of age, those with perianal disease at diagnosis and/or those who initially require steroids, might benefit from early aggressive intervention [37].

Mucosal Healing

Mucosal healing is a major therapeutic goal in clinical trials. Mounting evidence indicates that mucosal healing might be useful in clinical practice. Mucosal healing induced by long-term maintenance infliximab treatment is associated with a lower need for major abdominal surgeries [38]. In UC, mucosal healing may represent the ultimate therapeutic goal because inflammation is limited to mucosa. In CD, a transmural disease, mucosal healing may be considered as the minimum therapeutic goal.

When to Stop Anti-TNF Therapy?
It is well established that both adalimumab and infliximab maintenance therapy reduces hospitalizations and surgeries at 1 year in patients with CD [39, 40]. An observational study assessed the long-term outcome of infliximab in 614 consecutive CD patients. After a median follow-up of 55 months, a sustained clinical benefit was observed in almost two-thirds of primary responders to infliximab, supporting its long-term use beyond 1 year in IBD [41].

Should We Use Anti-TNF Therapy Alone or in Combination with Immunomodulators?
A way to reduce immunogenicity and loss of response to infliximab may be the use of concomitant immunomodulators. One study showed that patients in remission taking both infliximab and azathioprine for 6 months, who then stopped azathioprine, had no difference in outcome at 104 weeks compared with patients who continued the infliximab-azathioprine combination [42].

Definitive results of the SONIC trial comparing azathioprine alone versus infliximab alone versus combination therapy will allow more accurate conclusions on the concomitant use of thiopurine in association with infliximab in the long-term [43]. In contrast to the previous study by Van Assche et al. [42], preliminary analysis of the SONIC study suggested that combination therapy was more effective than azathioprine or infliximab alone in maintaining corticosteroid-free remission at week 26 [43]. However, the benefit of combination therapy after one year of therapy will require additional investigations.

The risk-benefit assessment of combination therapy, in lieu of recent reports of hepatosplenic T-cell lymphomas in young males receiving combination therapy [44] as well as the increased risk of opportunistic infections [31], has led to a re-evaluation of recommendations for concurrent immunomodulatory therapy with anti-TNF agents.

Taken together, the risk-benefit ratio of combination therapy should be carefully evaluated in each individual patient. The continuation of immunosuppressants beyond 6 months seems to offer no clear benefit over scheduled infliximab monotherapy alone and cannot be systematically recommended, especially in young males.

Conclusion

Infliximab represents the pinnacle of the therapeutic pyramid of IBD treatment at this point in time. However, this anti-TNF agent has several limitations. First, despite its widespread use in IBD, 20% of patients still require surgery [41]. Second, about 10% of patients are primary non-responders to infliximab and only one-third of IBD patients are in clinical remission at 1 year [4, 41]. Third, the annual risk of loss of response is 13% per patient-year [45]. Finally, infliximab treatment optimization with combination therapy can be considered, but this must be weighed against the increased risk of serious infections and perhaps lymphoma. These data underscore the urgent need to develop new dug classes.

Adalimumab and certolizumab are thought to be less immunogenic than infliximab and concomitant use of immunomodulators is not recommended for adalimumab and certolizumab. However, all anti-TNF agents have the potential for immunogenicity [7, 9, 17] and indirect comparison between studies suggests that all of these agents have broadly similar efficacy in maintaining clinical remission at 6 months in luminal CD [4]. In addition, treatment safety seems to be the same with all of these anti-TNF agents. No head-to-head trials have directly compared the safety or efficacy between anti-TNF agents. Pending these results, other parameters such as the mode of administration and cost-efficacy should be taken into account when considering the best strategy to use anti-TNF agents.

In the future, anti-TNF therapy may be useful for other indications in the treatment of IBD. Notably, infliximab may be more effective than placebo to prevent endoscopic and histological recurrence of CD after ileal resection [46].

References

1 Kollias G, Douni E, Kassiotis G, Kontoyiannis D: On the role of tumor necrosis factor and receptors in models of multiorgan failure, rheumatoid arthritis, multiple sclerosis and inflammatory bowel disease. Immunol Rev 1999;169:175–194.

2 MacDonald TT, Hutchings P, Choy MY, Murch S, Cooke A: Tumour necrosis factor-alpha and interferon-gamma production measured at the single cell level in normal and inflamed human intestine. Clin Exp Immunol 1990;81:301–305.

3 van Dullemen HM, van Deventer SJ, Hommes DW, Bijl HA, Jansen J, Tytgat GN, Woody J: Treatment of Crohn's disease with anti-tumor necrosis factor chimeric monoclonal antibody (cA2). Gastroenterology 1995;109:129–135.

4 Peyrin-Biroulet L, Deltenre P, de Suray N, Branche J, Sandborn WJ, Colombel JF: Efficacy and safety of tumor necrosis factor antagonists in Crohn's disease: meta-analysis of placebo-controlled trials. Clin Gastroenterol Hepatol 2008;6:644–653.

5 Cassinotti A, Travis S: Incidence and clinical significance of immunogenicity to infliximab in Crohn's disease: a critical systematic review. Inflamm Bowel Dis 2009;15:1264–1275.

6 Targan SR, Hanauer SB, van Deventer SJ, Mayer L, Present DH, Braakman T, DeWoody KL, Schaible TF, Rutgeerts PJ: A short-term study of chimeric monoclonal antibody cA2 to tumor necrosis factor alpha for Crohn's disease. Crohn's Disease cA2 Study Group. N Engl J Med 1997;337:1029–1035.

7 Hanauer SB, Sandborn WJ, Rutgeerts P, Fedorak RN, Lukas M, MacIntosh D, Panaccione R, Wolf D, Pollack P: Human anti-tumor necrosis factor monoclonal antibody (adalimumab) in Crohn's disease: the CLASSIC-I trial. Gastroenterology 2006;130: 323–333.

8 Schreiber S, Rutgeerts P, Fedorak RN, Khaliq-Kareemi M, Kamm MA, Boivin M, Bernstein CN, Staun M, Thomsen OO, Innes A: A randomized, placebo-controlled trial of certolizumab pegol (CDP870) for treatment of Crohn's disease. Gastroenterology 2005;129:807–818.

9 Sandborn WJ, Feagan BG, Stoinov S, Honiball PJ, Rutgeerts P, Mason D, Bloomfield R, Schreiber S: Certolizumab pegol for the treatment of Crohn's disease. N Engl J Med 2007;357:228–238.

10 Rutgeerts P, D'Haens G, Targan S, Vasiliauskas E, Hanauer SB, Present DH, Mayer L, Van Hogezand RA, Braakman T, DeWoody KL, Schaible TF, Van Deventer SJ: Efficacy and safety of retreatment with anti-tumor necrosis factor antibody (infliximab) to maintain remission in Crohn's disease. Gastroenterology 1999;117:761–769.

11 Hanauer SB, Feagan BG, Lichtenstein GR, Mayer LF, Schreiber S, Colombel JF, Rachmilewitz D, Wolf DC, Olson A, Bao W, Rutgeerts P: Maintenance infliximab for Crohn's disease: the ACCENT I randomised trial. Lancet 2002;359:1541–1549.

12 Hanauer SB, Wagner CL, Bala M, Mayer L, Travers S, Diamond RH, Olson A, Bao W, Rutgeerts P: Incidence and importance of antibody responses to infliximab after maintenance or episodic treatment in Crohn's disease. Clin Gastroenterol Hepatol 2004;2:542–553.
13 Rutgeerts P, Diamond RH, Bala M, Olson A, Lichtenstein GR, Bao W, Patel K, Wolf DC, Safdi M, Colombel JF, Lashner B, Hanauer SB: Scheduled maintenance treatment with infliximab is superior to episodic treatment for the healing of mucosal ulceration associated with Crohn's disease. Gastrointest Endosc 2006;63:433–442.
14 Rutgeerts P, Feagan BG, Lichtenstein GR, Mayer LF, Schreiber S, Colombel JF, Rachmilewitz D, Wolf DC, Olson A, Bao W, Hanauer SB: Comparison of scheduled and episodic treatment strategies of infliximab in Crohn's disease. Gastroenterology 2004; 126:402–413.
15 Colombel JF, Sandborn WJ, Rutgeerts P, Enns R, Hanauer SB, Panaccione R, Schreiber S, Byczkowski D, Li J, Kent JD, Pollack PF: Adalimumab for maintenance of clinical response and remission in patients with Crohn's disease: the CHARM trial. Gastroenterology 2007;132:52–65.
16 Colombel JF, Sandborn WJ, Rutgeerts P, Kamm MA, Yu AP, Wu EQ, Pollack PF, Lomax KG, Chao J, Mulani PM: Comparison of two adalimumab treatment schedule strategies for moderate-to-severe Crohn's disease: results from the CHARM trial. Am J Gastroenterol 2009;104:1170–1179.
17 Schreiber S, Khaliq-Kareemi M, Lawrance IC, Thomsen OO, Hanauer SB, McColm J, Bloomfield R, Sandborn WJ: Maintenance therapy with certolizumab pegol for Crohn's disease. N Engl J Med 2007;357:239–250.
18 Sands BE, Anderson FH, Bernstein CN, Chey WY, Feagan BG, Fedorak RN, Kamm MA, Korzenik JR, Lashner BA, Onken JE, Rachmilewitz D, Rutgeerts P, Wild G, Wolf DC, Marsters PA, Travers SB, Blank MA, van Deventer SJ: Infliximab maintenance therapy for fistulizing Crohn's disease. N Engl J Med 2004;350:876–885.
19 Present DH, Rutgeerts P, Targan S, Hanauer SB, Mayer L, van Hogezand RA, Podolsky DK, Sands BE, Braakman T, DeWoody KL, Schaible TF, van Deventer SJ: Infliximab for the treatment of fistulas in patients with Crohn's disease. N Engl J Med 1999;340:1398–1405.
20 Colombel JF, Schwartz DA, Sandborn WJ, Kamm MA, D'Haens G, Rutgeerts PJ, Enns RA, Panaccione R, Schreiber S, Li J, Kent JD, Lomax KG, Pollack PF: Adalimumab for the treatment of fistulas in patients with Crohn's disease. Gut 2009;58:940–948.
21 Rutgeerts P, Sandborn WJ, Feagan BG, Reinisch W, Olson A, Johanns J, Travers S, Rachmilewitz D, Hanauer SB, Lichtenstein GR, de Villiers WJ, Present D, Sands BE, Colombel JF: Infliximab for induction and maintenance therapy for ulcerative colitis. N Engl J Med 2005;353:2462–2476.
22 Sandborn WJ, Rutgeerts P, Enns R, Hanauer SB, Colombel JF, Panaccione R, D'Haens G, Li J, Rosenfeld MR, Kent JD, Pollack PF: Adalimumab induction therapy for Crohn disease previously treated with infliximab: a randomized trial. Ann Intern Med 2007;146:829–838.
23 Oussalah A, Babouri A, Chevaux JB, Stancu L, Trouilloud I, Bensenane M, Boucekkine T, Bigard MA, Peyrin-Biroulet L: Adalimumab for Crohn's disease with intolerance or lost response to infliximab: a 3-year single-centre experience. Aliment Pharmacol Ther 2009;29:416–423.
24 Peyrin-Biroulet L, Laclotte C, Roblin X, Bigard MA: Adalimumab induction therapy for ulcerative colitis with intolerance or lost response to infliximab: an open-label study. World J Gastroenterol 2007;13: 2328–2332.
25 Oussalah A, Laclotte C, Chevaux JB, Bensenane M, Babouri A, Serre AA, Boucekkine T, Roblin X, Bigard MA, Peyrin-Biroulet L: Long-term outcome of adalimumab therapy for ulcerative colitis with intolerance or lost response to infliximab: a single-centre experience. Aliment Pharmacol Ther 2008; 28:966–972.
26 Afif W, Leighton JA, Hanauer SB, Loftus EV Jr, Faubion WA, Pardi DS, Tremaine WJ, Kane SV, Bruining DH, Cohen RD, Rubin DT, Hanson KA, Sandborn WJ: Open-label study of adalimumab in patients with ulcerative colitis including those with prior loss of response or intolerance to infliximab. Inflamm Bowel Dis 2009, E-pub ahead of print.
27 Vermeire S, Abreu MT, D'Haens G, et al: Efficacy and safety of certolizumab pegol in patients with active Crohn's disease who previously lost response or were intolerant to infliximab: open-label induction preliminary results of the WELCOME study. Gastroenterology 2008;134:A67.
28 Colombel JF, Loftus EV Jr, Tremaine WJ, Egan LJ, Harmsen WS, Schleck CD, Zinsmeister AR, Sandborn WJ: The safety profile of infliximab in patients with Crohn's disease: the Mayo clinic experience in 500 patients. Gastroenterology 2004;126: 19–31.
29 Fidder H, Schnitzler F, Ferrante M, Noman M, Katsanos K, Segaert S, Henckaerts L, Van Assche G, Vermeire S, Rutgeerts P: Long-term safety of infliximab for the treatment of inflammatory bowel disease: a single-centre cohort study. Gut 2009;58: 501–508.

30 Lichtenstein GR, Feagan BG, Cohen RD, Salzberg BA, Diamond RH, Chen DM, Pritchard ML, Sandborn WJ: Serious infections and mortality in association with therapies for Crohn's disease: TREAT registry. Clin Gastroenterol Hepatol 2006;4:621–630.

31 Toruner M, Loftus EV Jr, Harmsen WS, Zinsmeister AR, Orenstein R, Sandborn WJ, Colombel JF, Egan LJ: Risk factors for opportunistic infections in patients with inflammatory bowel disease. Gastroenterology 2008;134:929–936.

32 Colombel JF, Sandborn WJ, Panaccione R, Robinson AM, Lau W, Li J, Cardoso AT: Adalimumab safety in global clinical trials of patients with Crohn's disease. Inflamm Bowel Dis 2009, E-pub ahead of print.

33 Siegel CA, Marden SM, Persing SH, Larson RJ, Sands BE: Risk of lymphoma associated with combination anti-tumor necrosis factor and immunomodulator therapy for the treatment of Crohn's disease: a meta-analysis. Clin Gastroenterol Hepatol 2009, E-pub ahead of print.

34 Peyrin-Biroulet L, Colombel JF, Sandborn WJ: Insufficient evidence to conclude whether anti-TNF therapy increases the risk of lymphoma in Crohn's disease. Clin Gastroenterol Hepatol 2009, E-pub ahead of print.

35 Lichtenstein GR, Diamond RH, Wagner CL, Fasanmade AA, Olson AD, Marano CW, Johanns J, Lang Y, Sandborn WJ: Clinical trials: benefits and risks of immunomodulators and maintenance infliximab for IBD-subgroup analyses across four randomized trials. Aliment Pharmacol Ther 2009;30:210–226.

36 D'Haens G, Baert F, van Assche G, Caenepeel P, Vergauwe P, Tuynman H, De Vos M, van Deventer S, Stitt L, Donner A, Vermeire S, Van de Mierop FJ, Coche JC, van der Woude J, Ochsenkuhn T, van Bodegraven AA, Van Hootegem PP, Lambrecht GL, Mana F, Rutgeerts P, Feagan BG, Hommes D: Early combined immunosuppression or conventional management in patients with newly diagnosed Crohn's disease: an open randomised trial. Lancet 2008;371:660–667.

37 Beaugerie L, Seksik P, Nion-Larmurier I, Gendre JP, Cosnes J: Predictors of Crohn's disease. Gastroenterology 2006;130:650–656.

38 Schnitzler F, Fidder H, Ferrante M, Noman M, Arijs I, Van Assche G, Hoffman I, Van Steen K, Vermeire S, Rutgeerts P: Mucosal healing predicts long-term outcome of maintenance therapy with infliximab in Crohn's disease. Inflamm Bowel Dis 2009, E-pub ahead of print.

39 Lichtenstein GR, Yan S, Bala M, Blank M, Sands BE. Infliximab maintenance treatment reduces hospitalizations, surgeries, and procedures in fistulizing Crohn's disease. Gastroenterology 2005;128:862–869.

40 Feagan BG, Panaccione R, Sandborn WJ, D'Haens GR, Schreiber S, Rutgeerts PJ, Loftus EV, Jr., Lomax KG, Yu AP, Wu EQ, Chao J, Mulani P: Effects of adalimumab therapy on incidence of hospitalization and surgery in Crohn's disease: results from the CHARM study. Gastroenterology 2008;135:1493–1499.

41 Schnitzler F, Fidder H, Ferrante M, Noman M, Arijs I, Van Assche G, Hoffman I, Van Steen K, Vermeire S, Rutgeerts P: Long-term outcome of treatment with infliximab in 614 patients with Crohn's disease: results from a single-centre cohort. Gut 2009;58:492–500.

42 Van Assche G, Magdelaine-Beuzelin C, D'Haens G, Baert F, Noman M, Vermeire S, Ternant D, Watier H, Paintaud G, Rutgeerts P: Withdrawal of immunosuppression in Crohn's disease treated with scheduled infliximab maintenance: a randomized trial. Gastroenterology 2008;134:1861–1868.

43 Colombel JF, Rutgeerts P, Reinisch W, et al: SONIC: A randomized, double-blind, controlled trial comparing infliximab and inflixmab plus azathioprine to azathioprine in patients with Crohn's disease naive to immunomodulators and biologic therapy. Gut 2008;57(suppl 2):A1.

44 Shale M, Kanfer E, Panaccione R, Ghosh S: Hepatosplenic T cell lymphoma in inflammatory bowel disease. Gut 2008;57:1639–1641.

45 Gisbert JP, Panes J: Loss of response and requirement of infliximab dose intensification in Crohn's disease: a review. Am J Gastroenterol 2009;104:760–767.

46 Regueiro M, Schraut W, Baidoo L, Kip KE, Sepulveda AR, Pesci M, Harrison J, Plevy SE: Infliximab prevents Crohn's disease recurrence after ileal resection. Gastroenterology 2009;136:441–450.

Laurent Peyrin-Biroulet
Inserm, U954 and Department of Hepato-Gastroenterology,
University Hospital of Nancy, Allee du Morvan
FR–54511 Vandoeuvre-lès-Nancy (France)
Tel. +33 3 8315 3631, Fax +33 3 8315 3633, E-Mail peyrin-biroulet@netcourrier.com

Mayerle J, Tilg H (eds): Clinical Update on Inflammatory Disorders of the Gastrointestinal Tract.
Front Gastrointest Res. Basel, Karger, 2010, vol 26, pp 108–117

Role of Epithelial Cells in Inflammatory Bowel Disease

Arthur Kaser

Department of Medicine II, Innsbruck Medical University, Innsbruck, Austria

Abstract

A single layer of intestinal epithelial cells (IECs) forms the barrier between the largest accumulation of microbes in the body and the host itself. Functional insight into the biology of IECs has revealed a profound impact of IEC function on mucosal homeostasis and immune function. Moreover, genome-wide association studies and candidate gene studies have substantially expanded our insight into the genetic underpinning of human inflammatory bowel disease. Several pathways regulated by products of genes that are genetically associated with inflammatory bowel disease, like the endoplasmic stress response (XBP1), autophagy (ATG16L1), and innate immune pathways (NOD2) profoundly affect IEC function, in particular Paneth cells. Paneth cells reside at the base of small intestinal crypts and secrete abundant amounts of antimicrobial peptides, and are thereby considered to substantially affect the composition of the microbial flora. Further evidence from the study of NFκB signaling in IECs revealed important insights into how IECs regulate mucosal innate and adaptive immunity via cross-talk with dendritic cells, macrophages and CD4+ T cells. These studies indicate that IECs may play a profound role in regulating the interaction between commensal and infectious microbes of the intestinal tract and the host, with consequent implications for infectious and inflammatory diseases of the gastrointestinal tract.

Intestinal epithelial cells (IECs) have long been considered an inert physical barrier between the largest accumulation of microbes in the body and the host itself. Apart from this barrier function, IECs had been thought to primarily mediate the absorption of nutrients from the diet. However, several recent studies have revealed an unexpected and profound impact of IEC function on mucosal homeostastis. Cre/lox technology and consequent cell type-specific deletion of genes in mouse models has greatly facilitated these mechanistic in vivo studies. In parallel, genetic association studies have yielded numerous novel genetic loci and genes associated with both forms of inflammatory bowel disease (IBD), Crohn's disease (CD) and ulcerative colitis (UC). Among the IBD-associated genes where functional insight could be obtained, several affect IEC function [1]. This article will focus on recent insight into

IEC biology in the context of inflammatory bowel disease and highlight the impact IECs have on inflammatory and immune mechanisms in the intestines.

Endoplasmic Reticulum Stress

IECs, in particular Paneth cells and goblet cells, are among the most highly secretory cells in the body. A critical mechanism required in highly secretory cell types is the unfolded protein response (UPR), which allows cells to cope with conditions of endoplasmic reticulum stress due to the accumulation of unfolded or misfolded proteins. Three partly interrelated pathways, PERK/Atf4, ATF6/ATF6f, and IRE1/XBP1, represent the proximal effectors of the UPR, with the latter being the evolutionarily most conserved [2]. Genetic deletion of *Xbp1* specifically in the IEC compartment results in spontaneous small intestinal enteritis with histological hallmarks resembling IBD, including crypt abscesses, leucocyte infiltration and frank ulceration [3]. Complete *Xbp1* deficiency in IECs results in apoptotic depletion of Paneth cells and a numeric reduction of goblet cells. This concurs with a substantial impairment in the handling of oral infection with a model pathogen, *Listeria monocytogenes*, with increased colony forming units of *Listeria* recovered from faeces, as well as an increased translocation of bacteria to the liver in epithelial XBP1-deficient mice [3]. Notably, deletion of only one *Xbp1* allele within the IEC compartment was sufficient to induce enteritis in a substantial proportion of mice, but did not deplete Paneth cells, which appeared grossly normal on HE stainings. However, in in vitro assays of bactericidal function of isolated crypts, *Xbp1*$^{+/-}$ crypts exhibited an intermediate phenotype between full bactericidal function in *Xbp1*$^{+/+}$ and absent bactericidal function in *Xbp1*$^{-/-}$ crypts [3].

XBP1 deficiency in IECs, presumably due to the accumulation of unfolded or misfolded proteins in the absence of XBP1-induced UPR target genes, results in substantial over-activation of upstream IRE1. Such substantial IRE1 over-activation (as deduced from analysis of XBP1 mRNA splicing status) is even present in the small intestinal epithelium when only one XBP1 allele is deleted, which is associated with a ~50% reduction in XBP1 mRNA expression [3]. Active IRE1 has dual functionality in that it activates XBP1 via an unconventional splicing mechanism, and it recruits TRAF2, which ultimately leads to phosphorylation of JNK, a key inflammatory signal transducer [4]. Indeed, intestinal XBP1 deficiency is associated with increased JNK phosphorylation in the epithelium. Along similar lines, silencing of XBP1 mRNA expression in an IEC line results in increased JNK phosphorylation compared to control-silenced IECs upon stimulation with TNF-α or flagellin [3], a TLR5 ligand and the dominant antigen in Crohn's disease [5]. These data suggest that XBP1 within the intestinal epithelium might regulate 2 key aspects of IBD, namely the intestinal microbial flora (via Paneth cell function), and the propensity of the epithelium to respond to external stimuli, like inflammatory cytokines or TLR ligands present in abundance within the microbial flora. To put it differently, XBP1 and hence the endoplasmic reticulum (ER) stress response regulates the inflammatory tone of the intestinal epithelium, and impairment of the ER stress response due to decreased XBP1 function lets the intestinal epithelium react more vigorously to cytokines and bacterial-derived 'patterns' commonly present within the intestinal mucosa and the intestinal microbiota, respectively. This model proposes that cell-specific impairment of the UPR might result in organ-specific inflammation as observed in the intestine of mice with one or both *Xbp1* alleles deleted within the epithelium [3].

Since 3 earlier independent studies had suggested linkage of a locus close to the human XBP1 gene on chromosome 22 with both forms of IBD, CD and UC, a candidate gene study was performed in an index cohort and 2 replication cohorts of German patients and controls [3]. Indeed, genotyping of a HapMap-selected set of single nucleotide polymorphisms (SNPs) covering a 100-kb stretch around the *XBP1* locus revealed several significantly associated SNPs, establishing the *XBP1* locus as a novel risk locus for both CD and UC. Of note was the low degree of linkage disequilibrium across the *XBP1* locus as estimated by the r^2 metric, which suggested a complex genetic architecture of the locus and raised the possibility that rare rather than common variants of *XBP1* might be associated with disease. Indeed, deep sequencing of the *XBP1* gene in 1,000 IBD patients and controls revealed 3-fold more 'rare' SNPs in CD and UC patients compared to controls [3]. Among them were several coding, non-synonymous SNPs (nsSNPs). To reveal their functionality, 3 nsSNPs (2 which were only found in IBD patients, but not controls, and 1 that was present at equal frequency in IBD patients and controls alike) were selected and engineered into XBP1 cDNA expression vectors. Transfection into a small intestinal IEC line as well as reconstitution of $Xbp1^{-/-}$ mouse embryonic fibroblasts revealed decreased transactivation capacity of UPR target genes by the IBD-associated *XBP1* variants compared to wild-type *XBP1* [3]. In contrast, the nsSNP *XBP1* variant not associated with IBD did not exhibit any alteration in these assays, and was therefore predicted to be functionally silent [3]. These data suggested that multiple rare ('private') variants might account for the positional signal at the *XBP1* locus and that at least some of these variants could act through decreased transactivation capacity of *XBP1*, indicative that the mechanistic insight gained from the mouse model might be applicable to our understanding of human IBD-associated *XBP1* variants. Of further note is that human colonic and ileal biopsy samples obtained from CD and UC patients exhibited evidence of ER stress in that grp78 expression as well as XBP1 splicing is increased compared to healthy controls [3]. These data are corroborated by 2 further studies that also reported evidence of increased grp78 expression and XBP1 splicing, respectively, in IECs obtained from IBD patients [6, 7]. While genetic polymorphisms in *XBP1* might be infrequent (compared to NOD2, ATG16L1 and IL23R) as the underlying genetic susceptibility factor for IBD, these latter data might suggest that increased ER stress secondary to other environmental or genetic factors might also be an important perpetuator of inflammation in the intestine by the mechanisms elucidated in the epithelial-deficient *Xbp1* mouse model, and therefore might represent a potentially attractive pharmacological target at the inner surface of the intestines [8]. In this context, it is notable that in a murine N-ethyl-N-nitrosourea mutagenesis experiment, 2 independent missense mutations in the *Muc2* gene were identified as causative for the spontaneous development of a UC-like phenotype in 2 mouse models called *Winnie* and *Eeyore* [7]. Mutant mice exhibited aberrant MUC2 biosynthesis along with ultrastructural and biochemical evidence of ER stress and activation of the UPR in mucin-producing goblet cells, and expression of mutant MUC2 oligomerization domains induced ER stress in vitro [7]. Moreover, a similar accumulation of non-glycoslyated MUC2 precursors was reported in goblet cells of UC patients along with ultrastructural and biochemical evidence of ER stress even in non-inflamed intestinal tissue, suggesting that perturbations of the UPR could represent a broader theme for the mechanistic underpinning of IBD and UC in particular [7]. It is noteworthy that the intestinal epithelium is the only cell type that expresses 2 highly homologous forms of IRE1, ubiquitously expressed IRE1α and IEC-specific IRE1β, which could be interpreted as pointing toward a critical requirement of this pathway in IECs [9]. In fact, mice deficient in IRE1β exhibit increased sensitivity to dextran sodium sulphate (DSS) colitis [9], a phenotype also observed in mice with IEC-specific *Xbp1* deletion [3]. However, *Ire1β*$^{-/-}$ *(Ern2)* mice do not develop spontaneous intestinal inflammation [9]. It might be speculated that the genetic absence of IRE1β and hence inadequate activation of downstream effectors like XBP1 might lead to over-activation of IRE1α in IECs similar to XBP1-deficient mice, suggesting a converging mechanism of XBP1 and IRE1β deficiency in IECs leading to an increased inflammatory tone and increased susceptibility to DSS colitis.

Autophagy

Genome-wide association studies have identified several genes associated with autophagy as risk factors for CD [10–13]. This is particularly exciting, as autophagy has not previously been linked to the pathogenesis of IBD, and hence this discovery highlighted the potency of such association studies to generate novel unbiased hypothesis on the mechanistic basis of IBD. The first association of an autophagy gene, *ATG16L1*, with CD was initially discovered as a result of a genome-wide study on nsSNPs [10]. In fact, among the more than 30 genetic loci linked to IBD, *ATG16L1*, *NOD2* and *IL23R* polymorphisms are those most strongly associated with CD *(ATG16L1, NOD2)*, and CD + UC *(IL23R)* [12, 14]. In addition to *ATG16L1*, two more autophagy genes have been associated with IBD: *IRGM* and *LRRK2* [11, 12]. First insights into possible mechanisms on how ATG16L1 might contribute to the development of CD were recently reported. Genetically engineered mice expressing a hypomorphic variant of *Atg16l1* or mice with deletion of another autophagy gene, *Atg5*, selectively in the intestinal epithelium exhibited notable alterations of

Paneth cells in the terminal ileum [15]. Specifically, *Atg16l1* hypomorphic Paneth cells presented with morphological alterations in the granule exocytosis pathway, and transcriptional analysis revealed increased expression of genes in the peroxisome proliferator-activated receptor signalling and lipid metabolism, of acute phase reactants and 2 adipocytokines, leptin and adiponectin [15]. Other epithelial cell types apart from Paneth cells showed normal morphology, suggesting that Paneth cells might be particularly sensitive to disruption of autophagic pathways. Interestingly, despite these morphological changes in Paneth cell granules, *Atg16l1* hypomorphic mice were indistinguishable from wild-type mice in the translocation of the orally infected model pathogen *Listeria monocytogenes* [15]. However, notably, CD patients homozygous for the *ATG16L1* variant exhibited structural alterations in their Paneth cell granules similar to those observed in the hypomorphic *Atg16l1* mouse model, and they also showed increased expression of leptin [15]. Apart from regulating the secretory apparatus of Paneth cells, ATG16L1 has also been reported to regulate endotoxin-induced inflammasome activation and hence the secretion of IL-1β and IL-18 in myeloid cells [16]. Mice with a haematopoietic cell-specific deletion of *Atg16l1* apparently did not develop spontaneous disease, but were particularly sensitive to DSS colitis, which was alleviated by administration of anti-IL-1β and anti-IL-18 antibodies [16]. These data indicate that hypomorphic function of ATG16L1 might impede Paneth cell biology, and at the same time increase the inflammatory reactivity of the myeloid compartment. This is reminiscent of NOD2 function, which will be discussed in the next paragraph.

NOD2

The discovery of an association of polymorphisms in the pattern recognition gene *NOD2* and CD in 2001 was not only the first locus firmly associated with CD, but in fact the first time that polymorphisms in a single gene could be linked to the genetic basis of a polygenic disease [17–19]. Surprisingly, despite the early identification of associated nsSNPs that affect *NOD2* molecular function, a comprehensive model of disease pathogenesis based on altered NOD2 function has still not yet evolved, which might in part be due to the fact that neither *Nod2*$^{-/-}$ nor mice with a knock-in of the disease-associated major *NOD2* variant develop spontaneous disease [20, 21]. However, several lines of evidence indicate that altered NOD2 function might have substantial effects on myeloid cell and Paneth cell function. Specifically, *Nod2*$^{-/-}$ mice exhibit decreased expression of specific α-defensins (called cryptdins in mice) in Paneth cells, which is associated with a significant increased in the translocation of orally infected *Listeria monocytogenes* to the liver and spleen [20]. Similarly and compatible with the mouse model data, patients with ileal CD and the 3020insC *NOD2* frameshift mutation exhibit decreased expression of human α-defensin 5 and 6 [22]. In addition to these effects of NOD2 on Paneth cell biology, NOD2 also prominently affects inflammatory pathways. Engineering the 3020insC polymorphism (2939insC in mice) into the germ-line of mice results in elevated NFκB activation in myeloid cells upon stimulation with the NOD2 ligands muramyl dipeptide, along with more efficient processing and secretion of IL-1β [21]. Moreover, these mice exhibit an increased severity of colitis in the DSS model, which is alleviated by administration of recombinant IL-1 receptor antagonist (IL-1Ra) [21]. Unfortunately, the impact of the 2939insC mutation on α-defensin expression in Paneth cells and possible additional consequences on IEC biology have not been investigated. However, it is important to note that important species differences between mouse *Nod2*^{2939insC} and human *NOD2*^{3020insC} might be present, complicating the interpretation of these results, with human *NOD2*^{3020insC} exhibiting active suppression of the transcription of the anti-inflammatory cytokine IL-10 [23]. Another set of experimental evidence on NOD2 function showed that NOD2 inhibits TLR2-dependent NFκB activation [24–26]. Along these lines, genetic *Nod2* deficiency or IBD-associated *NOD2* variants exhibited increased NFκB signalling upon TLR2 stimulation [24–26]. In summary, these data show that NOD2 might affect multiple pathways ultimately contributing to the development of intestinal inflammation, with prominent effects on Paneth cell biology as well as inflammatory pathways in myeloid cells. This duality of effects on the epithelial cell compartment as well as on pro-inflammatory pathways is in some way reminiscent of the aforementioned functional properties of XBP1 and ATG16L1.

Organic Cation Transporters

Apart from *XBP1*, *ATG16L1*, and *NOD2*, with their ties to intestinal epithelial cell biology, other genetic loci associated with IBD have also been speculated to contribute to IBD pathophysiology through the IEC compartment. The IBD5 locus on chromosome 5q31 includes genes encoding the organic cation transporters OCTN1 and 2, which transport carnitine. A 250-kb haplotype block at the IBD5 locus has been associated with CD, and associated polymorphisms in OCTN1 (*SLC22A4*) and OCTN2 (*SLC22A5*) have been reported to cause decreased transporter activity and altered promoter activity, respectively [27–29]. OCTN1 and 2 show prominent expression specifically in the IEC layer of the colon [29], and IECs exhibit high expression and activity of enzymes of the β oxidation pathway [30]. OCTN2-deficient juvenile visceral steatosis mice develop inflammatory changes ranging from intestinal villous atrophy to ulcer formation and perforation, with the inflammatory infiltrate being characterized by lymphocyte and macrophage infiltration [30]. These data suggest that fatty acid oxidation, which is regulated by carnitine-dependent entry of long-chain fatty acids into the mitochondrial matrix, might be an important pathway for CD. However, it should also be cautioned that while the IBD5 locus is firmly established and widely replicated as a CD risk locus, there remains some uncertainty as to the causal variant within this region [reviewed in 31]. While interferon regulatory factor-1 *(IRF1)* and prolyle-4 hydroxylase *(P4HA2)* are equally likely candidates at the IBD5 locus as *SLC22A4* and *SLC22A5*, as deduced from genetic association analysis [31], the functional evidence for OCTN2 may render it a particularly likely candidate [30]. This view is further corroborated by expression quantitative trait locus analysis, which revealed that CD-associated SNPs at the IBD5 locus are the same variants most associated with (decreased) *SLC22A5* expression [12].

NFκB

While the IEC-centred mechanisms discussed in the previous paragraphs are immediately related to genes associated with IBD as risk factors of disease, studies of the NFκB pathway within IECs have elegantly revealed that IECs tightly control innate and adaptive immune mechanisms within the mucosa. Specifically, inhibitor of κB kinase 2 (IKK2, *Ikbkb*), upstream of NFκB, is a critical regulator of dendritic cell and CD4+ T cell function [32]. Mice with an IEC-specific deletion of IKK2 fail to eradicate an infection with the gut-dwelling worm *Trichuris muris*, which is associated with a failure to develop protective antigen-specific CD4+ Th2 immunity. Instead of a protective Th2 immune response, these mice exhibit increased IL-12/23 p40 and TNF-α expression and develop an inflammatory CD4+ T cell response characterized by increased expression of IFN-γ and IL-17, and consequently develop severe intestinal inflammation [32]. Lack of IEC-derived thymic stromal lymphopoietin acting on dendritic cells is critical in leading to these abnormalities in mice lacking IKK2 expression in IECs [33]. IEC-derived thymic stromal lymphopoietin had earlier been proposed from in vitro studies as an IEC-derived factor regulating dendritic cell function and consequent development of a CD4+ T cell response [34]. These data not only point toward an intriguing role of IECs in directing innate and adaptive immune mechanisms in the mucosa, but also highlight profound differences in the consequences of NFκB signalling in IECs compared to myeloid cells. While genetic IKK2 deletion in IECs leads to increased severity of the active phase of experimental colitis induced by the chemical barrier-disruptive toxin DSS, IKK2 deletion during the chronic phase of DSS colitis does not impact on disease severity [35]. Similarly, IKK2 deletion in IECs does not affect the severity of spontaneous colitis developing in $Il10^{-/-}$ mice, which is in marked contrast to myeloid-specific deletion of IKK2, which significantly attenuates disease [35]. These data highlight the pre-eminent importance of context- and tissue dependency of NFκB signalling. IKK2 activates the canonical NFκB pathway, while IKK1 predominantly activates the alternative pathway [36]. Both proteins form a complex with NEMO (NFκB essential modulator, IKK-γ) [36]. Deletion of either NEMO, or both IKK1 and IKK2, specifically in IECs results in apoptotic loss of IECs, translocation of bacteria to the mucosa, and consequent development of severe spontaneous colitis [37]. NEMO deletion in IECs sensitizes IECs to TNF-induced apoptosis, and *Tnfr1*$^{-/-}$ mice attenuates intestinal inflammation in mice with IEC-specific deletion of NEMO [37]. Overall, these studies point towards a critical role of NFκB signalling in IECs and consequent regulation of inflammatory and immune mechanisms in the mucosa.

Conclusion

IECs have emerged as critical immunological determinants of inflammatory and immune functions in the intestine. Their location at the interface in-between the intestinal microbiota and the mammalian host implies an important role in host-microbiota mutualism, which is an important aspect in IBD in particular [38]. Moreover, the IECs' location at the inner surface of the intestine makes them attractive targets for pharmacological compounds designed to affect IEC-specific pathways related to the development of IBD.

References

1 Kaser A, Blumberg RS: Paneth cells and inflammation dance together in Crohn's disease. Cell Res 2008;18:1160–1162.
2 Ron D, Walter P: Signal integration in the endoplasmic reticulum unfolded protein response. Nat Rev Mol Cell Biol 2007;8:519–529.
3 Kaser A, Lee AH, Franke A, Glickman JN, Zeissig S, Tilg H, Nieuwenhuis EE, Higgins DE, Schreiber S, Glimcher LH, Blumberg RS: XBP1 links ER stress to intestinal inflammation and confers genetic risk for human inflammatory bowel disease. Cell 2008;134: 743–756.
4 Todd DJ, Lee AH, Glimcher LH: The endoplasmic reticulum stress response in immunity and autoimmunity. Nat Rev Immunol 2008;8:663–674.
5 Lodes MJ, Cong Y, Elson CO, Mohamath R, Landers CJ, Targan SR, Fort M, Hershberg RM: Bacterial flagellin is a dominant antigen in Crohn disease. J Clin Invest 2004;113:1296–1306.
6 Shkoda A, Ruiz PA, Daniel H, Kim SC, Rogler G, Sartor RB, Haller D: Interleukin-10 blocked endoplasmic reticulum stress in intestinal epithelial cells: impact on chronic inflammation. Gastroenterology 2007;132:190–207.
7 Heazlewood CK, Cook MC, Eri R, Price GR, Tauro SB, Taupin D, Thornton DJ, Png CW, Crockford TL, Cornall RJ, Adams R, Kato M, Nelms KA, Hong NA, Florin TH, Goodnow CC, McGuckin MA. Aberrant mucin assembly in mice causes endoplasmic reticulum stress and spontaneous inflammation resembling ulcerative colitis. PLoS Med 2008;5:e54.
8 Kaser A, Blumberg RS: Endoplasmic reticulum stress in the intestinal epithelium and inflammatory bowel disease. Semin Immunol 2009;21:156–63.
9 Bertolotti A, Wang X, Novoa I, Jungreis R, Schlessinger K, Cho JH, West AB, Ron D: Increased sensitivity to dextran sodium sulfate colitis in IRE1beta-deficient mice. J Clin Invest 2001;107:585–593.
10 Hampe J, Franke A, Rosenstiel P, Till A, Teuber M, Huse K, Albrecht M, Mayr G, De La Vega FM, Briggs J, Gunther S, Prescott NJ, Onnie CM, Hasler R, Sipos B, Folsch UR, Lengauer T, Platzer M, Mathew CG, Krawczak M, Schreiber S: A genome-wide association scan of nonsynonymous SNPs identifies a susceptibility variant for Crohn disease in ATG16L1. Nat Genet 2007;39:207–211.
11 The Wellcome Trust Case Control Consortium: Genome-wide association study of 14,000 cases of seven common diseases and 3,000 shared controls. Nature 2007;447:661–678.

12 Barrett JC, Hansoul S, Nicolae DL, Cho JH, Duerr RH, Rioux JD, Brant SR, Silverberg MS, Taylor KD, Barmada MM, Bitton A, Dassopoulos T, Datta LW, Green T, Griffiths AM, Kistner EO, Murtha MT, Regueiro MD, Rotter JI, Schumm LP, Steinhart AH, Targan SR, Xavier RJ, Libioulle C, Sandor C, Lathrop M, Belaiche J, Dewit O, Gut I, Heath S, Laukens D, Mni M, Rutgeerts P, Van Gossum A, Zelenika D, Franchimont D, Hugot JP, de Vos M, Vermeire S, Louis E, Cardon LR, Anderson CA, Drummond H, Nimmo E, Ahmad T, Prescott NJ, Onnie CM, Fisher SA, Marchini J, Ghori J, Bumpstead S, Gwilliam R, Tremelling M, Deloukas P, Mansfield J, Jewell D, Satsangi J, Mathew CG, Parkes M, Georges M, Daly MJ: Genome-wide association defines more than 30 distinct susceptibility loci for Crohn's disease. Nat Genet 2008;40:955–962.

13 Rioux JD, Xavier RJ, Taylor KD, Silverberg MS, Goyette P, Huett A, Green T, Kuballa P, Barmada MM, Datta LW, Shugart YY, Griffiths AM, Targan SR, Ippoliti AF, Bernard EJ, Mei L, Nicolae DL, Regueiro M, Schumm LP, Steinhart AH, Rotter JI, Duerr RH, Cho JH, Daly MJ, Brant SR: Genome-wide association study identifies new susceptibility loci for Crohn disease and implicates autophagy in disease pathogenesis. Nat Genet 2007;39:596–604.

14 Duerr RH, Taylor KD, Brant SR, Rioux JD, Silverberg MS, Daly MJ, Steinhart AH, Abraham C, Regueiro M, Griffiths A, Dassopoulos T, Bitton A, Yang H, Targan S, Datta LW, Kistner EO, Schumm LP, Lee AT, Gregersen PK, Barmada MM, Rotter JI, Nicolae DL, Cho JH: A genome-wide association study identifies IL23R as an inflammatory bowel disease gene. Science 2006;314:1461–1463.

15 Cadwell K, Liu JY, Brown SL, Miyoshi H, Loh J, Lennerz JK, Kishi C, Kc W, Carrero JA, Hunt S, Stone CD, Brunt EM, Xavier RJ, Sleckman BP, Li E, Mizushima N, Stappenbeck TS, Virgin Iv HW: A key role for autophagy and the autophagy gene Atg16l1 in mouse and human intestinal Paneth cells. Nature 2008;456:259–263.

16 Saitoh T, Fujita N, Jang MH, Uematsu S, Yang BG, Satoh T, Omori H, Noda T, Yamamoto N, Komatsu M, Tanaka K, Kawai T, Tsujimura T, Takeuchi O, Yoshimori T, Akira S: Loss of the autophagy protein Atg16L1 enhances endotoxin-induced IL-1beta production. Nature 2008;456:264–268.

17 Ogura Y, Bonen DK, Inohara N, Nicolae DL, Chen FF, Ramos R, Britton H, Moran T, Karaliuskas R, Duerr RH, Achkar JP, Brant SR, Bayless TM, Kirschner BS, Hanauer SB, Nunez G, Cho JH: A frameshift mutation in NOD2 associated with susceptibility to Crohn's disease. Nature 2001;411:603–606.

18 Hugot JP, Chamaillard M, Zouali H, Lesage S, Cezard JP, Belaiche J, Almer S, Tysk C, O'Morain CA, Gassull M, Binder V, Finkel Y, Cortot A, Modigliani R, Laurent-Puig P, Gower-Rousseau C, Macry J, Colombel JF, Sahbatou M, Thomas G: Association of NOD2 leucine-rich repeat variants with susceptibility to Crohn's disease. Nature 2001;411:599–603.

19 Hampe J, Cuthbert A, Croucher PJ, Mirza MM, Mascheretti S, Fisher S, Frenzel H, King K, Hasselmeyer A, MacPherson AJ, Bridger S, van Deventer S, Forbes A, Nikolaus S, Lennard-Jones JE, Foelsch UR, Krawczak M, Lewis C, Schreiber S, Mathew CG: Association between insertion mutation in NOD2 gene and Crohn's disease in German and British populations. Lancet 2001;357:1925–1928.

20 Kobayashi KS, Chamaillard M, Ogura Y, Henegariu O, Inohara N, Nunez G, Flavell RA: Nod2-dependent regulation of innate and adaptive immunity in the intestinal tract. Science 2005;307:731–734.

21 Maeda S, Hsu LC, Liu H, Bankston LA, Iimura M, Kagnoff MF, Eckmann L, Karin M: Nod2 mutation in Crohn's disease potentiates NF-kappaB activity and IL-1beta processing. Science 2005;307:734–738.

22 Wehkamp J, Salzman NH, Porter E, Nuding S, Weichenthal M, Petras RE, Shen B, Schaeffeler E, Schwab M, Linzmeier R, Feathers RW, Chu H, Lima H Jr, Fellermann K, Ganz T, Stange EF, Bevins CL: Reduced Paneth cell alpha-defensins in ileal Crohn's disease. Proc Natl Acad Sci USA 2005;102:18129–18134.

23 Noguchi E, Homma Y, Kang X, Netea MG, Ma X: A Crohn's disease-associated NOD2 mutation suppresses transcription of human IL10 by inhibiting activity of the nuclear ribonucleoprotein hnRNP-A1. Nat Immunol 2009;10:471–479.

24 Watanabe T, Kitani A, Murray PJ, Strober W: NOD2 is a negative regulator of Toll-like receptor 2-mediated T helper type 1 responses. Nat Immunol 2004;5:800–808.

25 Watanabe T, Asano N, Murray PJ, Ozato K, Tailor P, Fuss IJ, Kitani A, Strober W: Muramyl dipeptide activation of nucleotide-binding oligomerization domain 2 protects mice from experimental colitis. J Clin Invest 2008;118:545–559.

26 Watanabe T, Kitani A, Murray PJ, Wakatsuki Y, Fuss IJ, Strober W: Nucleotide binding oligomerization domain 2 deficiency leads to dysregulated TLR2 signaling and induction of antigen-specific colitis. Immunity 2006;25:473–485.

27 Rioux JD, Silverberg MS, Daly MJ, Steinhart AH, McLeod RS, Griffiths AM, Green T, Brettin TS, Stone V, Bull SB, Bitton A, Williams CN, Greenberg GR, Cohen Z, Lander ES, Hudson TJ, Siminovitch KA: Genomewide search in Canadian families with inflammatory bowel disease reveals two novel susceptibility loci. Am J Hum Genet 2000;66:1863–1870.

28 Rioux JD, Daly MJ, Silverberg MS, Lindblad K, Steinhart H, Cohen Z, Delmonte T, Kocher K, Miller K, Guschwan S, Kulbokas EJ, O'Leary S, Winchester E, Dewar K, Green T, Stone V, Chow C, Cohen A, Langelier D, Lapointe G, Gaudet D, Faith J, Branco N, Bull SB, McLeod RS, Griffiths AM, Bitton A, Greenberg GR, Lander ES, Siminovitch KA, Hudson TJ: Genetic variation in the 5q31 cytokine gene cluster confers susceptibility to Crohn disease. Nat Genet 2001;29:223–228.

29 Peltekova VD, Wintle RF, Rubin LA, Amos CI, Huang Q, Gu X, Newman B, Van Oene M, Cescon D, Greenberg G, Griffiths AM, George-Hyslop PH, Siminovitch KA: Functional variants of OCTN cation transporter genes are associated with Crohn disease. Nat Genet 2004;36:471–475.

30 Shekhawat PS, Srinivas SR, Matern D, Bennett MJ, Boriack R, George V, Xu H, Prasad PD, Roon P, Ganapathy V: Spontaneous development of intestinal and colonic atrophy and inflammation in the carnitine-deficient jvs (OCTN2(-/-)) mice. Mol Genet Metab 2007;92:315–324.

31 Silverberg MS: OCTNs: will the real IBD5 gene please stand up? World J Gastroenterol 2006;12: 3678–3681.

32 Zaph C, Troy AE, Taylor BC, Berman-Booty LD, Guild KJ, Du Y, Yost EA, Gruber AD, May MJ, Greten FR, Eckmann L, Karin M, Artis D: Epithelial-cell-intrinsic IKK-beta expression regulates intestinal immune homeostasis. Nature 2007;446:552–556.

33 Taylor BC, Zaph C, Troy AE, Du Y, Guild KJ, Comeau MR, Artis D: TSLP regulates intestinal immunity and inflammation in mouse models of helminth infection and colitis. J Exp Med 2009;206: 655–667.

34 Rimoldi M, Chieppa M, Salucci V, Avogadri F, Sonzogni A, Sampietro GM, Nespoli A, Viale G, Allavena P, Rescigno M: Intestinal immune homeostasis is regulated by the crosstalk between epithelial cells and dendritic cells. Nat Immunol 2005;6:507–514.

35 Eckmann L, Nebelsiek T, Fingerle AA, Dann SM, Mages J, Lang R, Robine S, Kagnoff MF, Schmid RM, Karin M, Arkan MC, Greten FR: Opposing functions of IKKbeta during acute and chronic intestinal inflammation. Proc Natl Acad Sci USA 2008;105:15058–15063.

36 Vallabhapurapu S, Karin M: Regulation and function of NF-kappaB transcription factors in the immune system. Annu Rev Immunol 2009;27:693–733.

37 Nenci A, Becker C, Wullaert A, Gareus R, van Loo G, Danese S, Huth M, Nikolaev A, Neufert C, Madison B, Gumucio D, Neurath MF, Pasparakis M: Epithelial NEMO links innate immunity to chronic intestinal inflammation. Nature 2007;446:557–561.

38 Frank DN, St Amand AL, Feldman RA, Boedeker EC, Harpaz N, Pace NR: Molecular-phylogenetic characterization of microbial community imbalances in human inflammatory bowel diseases. Proc Natl Acad Sci USA 2007;104:13780–13785.

Arthur Kaser, MD
Department of Medicine II (Gastroenterology and Hepatology)
Innsbruck Medical University
Anichstrasse 35, AT–6020 Innsbruck (Austria)
Tel. +43 512 504 0, Fax +43 512 504 6725616, E-Mail arthur.kaser@i-med.ac.at

Gastroenterology

Mayerle J, Tilg H (eds): Clinical Update on Inflammatory Disorders of the Gastrointestinal Tract.
Front Gastrointest Res. Basel, Karger, 2010, vol 26, pp 118–125

GI Immune Response in Functional GI Disorders

Jan Tack · Sébastien Kindt

Department of Pathophysiology, Gastroenterology Section, University of Leuven, Leuven, Belgium

Abstract

Functional gastrointestinal disorders (FGIDs) are characterized by gastrointestinal symptoms in the absence of underlying organic disease. The pathophysiology underlying these disorders is poorly understood. Several recent studies have reported signs of low-grade inflammation on biopsies from the gastrointestinal tract in patients with FGIDs. This has mainly been studied in the irritable bowel syndrome, but more recent reports indicate similar findings in functional dyspepsia. There are also emerging data of systemic immune activation in at least a subset of patients with FGIDs, especially in cases with a post-infections origin. The contribution of immune activation to symptom generation and its role as a therapeutic target remain to be elucidated.

Introduction

The functional gastrointestinal disorders (FGIDs), are defined by the presence of gastrointestinal symptoms in the absence of an organic disease that readily explains their origin [1]. FGIDs are the most common disorders seen in gastrointestinal practice, where they have been estimated to account for more than 40% of patient visits [1, 2]. According to the Rome consensus, FGIDs are subdivided into specific syndromes based on the part of the gastrointestinal tract where symptoms are thought to originate, and on the specific symptom patterns [3]. The 2 most common FGIDs are functional dyspepsia (FD), a functional gastroduodenal disorder, and the irritable bowel syndrome (IBS), a functional bowel disorder [4, 5]. The pathophysiological determinants of FGIDs have been related to abnormal gastrointestinal motility, visceral hypersensitivity, mucosal immune alterations, and brain-gut dysregulation [6, 7].

Presence of Inflammation

Inflammation has been shown to have an impact in different diseases of the gastrointestinal tract. This is obvious in cases of gastrointestinal infections and inflammatory bowel disease. Although not suspected initially, a mild type of inflammation has also been demonstrated in patients with FGIDs, where routine investigations identify no apparent pathology. This has been best established for IBS, and most of the available studies focused on the lower gastrointestinal tract. Some more recent reports also indicate the presence of inflammatory changes in functional or motor disorders of other parts of the gastrointestinal tract, such as the stomach and small intestine.

Irritable Bowel Syndrome

Diagnostic criteria for IBS, based on the Rome III consensus [6], include recurrent abdominal pain or discomfort at least 3 days per month in the last 3 months associated with 2 out of 3 of the following: improvement with defecation, onset associated with a change in stool frequency, onset associated with a change in the appearance of stool. Although different putative pathophysiological mechanisms have been proposed, such as psychological factors, brain-gut axis dysfunction and visceral hypersensitivity, the exact mechanism remains unknown. There is emerging evidence in favour of a possible role of the immune system or a prior inflammatory event in at least a subset of IBS patients.

Epidemiological Evidence
As far back as 1962, Chaudhary and Truelove [8] had already reported that 34 out of 130 IBS patients dated the onset of their symptoms to a prior bacillary or amebic dysentery. Ever since, a higher incidence of bacterial or parasitic gastrointestinal infection preceding the onset of IBS has been indicated in retrospective [9, 10] as well as in prospective studies [11–15]. In one study a viral pathogen, a norovirus, preceded the onset of IBS symptomatology [15].

Immunological Changes in IBS
Other authors focused on alterations in biopsies from IBS patients. Three months after an acute gastroenteritis, the number of inflammatory cells (polymorphonuclear cells) remained elevated in rectal biopsies from patients who developed IBS (post-infectious IBS; PI-IBS), as compared to subjects with an uneventful recovery from prior gastroenteritis [16]. Higher counts of intra-epithelial lymphocytes were demonstrated in PI-IBS following *Campylobacter* enteritis [17]. Similarly, the number of intra-epithelial lymphocytes and enterochromaffin cells was increased in PI-IBS compared to controls 4 months after the acute episode [18]. An increased frequency of peripheral blood CD4+ and CD8+ cells expressing the gut homing receptor integrin β7 was found in IBS

compared to controls [19]. In the same study an augmented number of lamina propria CD8+ cells was detected in the ascending colon in IBS. Finally, Barbara et al. [20] pointed to the presence of an increased number of mast cells, an increased proportion of degranulating mast cells and more mast cells in close vicinity to nerve endings in IBS patients. The latter was even significantly correlated with severity and intensity of abdominal pain or discomfort [20]. Using the mucosal patch technique, an increased mucosal myeloperoxidase activity was demonstrated in IBS as compared to healthy individuals, although this elevation was less than in active ulcerative colitis [21].

Apart from alterations in local cellular immunity, involvement of different cytokines in susceptibility for IBS has also been suggested. Gwee et al. [22] reported a higher expression of IL-1β mRNA in rectal biopsies from PI-IBS patients 3 months after an episode of gastroenteritis. Single nucleotide polymorphisms in genes controlling the expression of different cytokines have been implicated, although the published series are relatively small and await confirmation. One study reported a higher prevalence of the low-producer genotype for the anti-inflammatory cytokine IL-10 in IBS patients [23]. This was not confirmed in another study, but a higher prevalence of the high-producer genotype of the pro-inflammatory TNF-α cytokine was reported [24].

Alterations in circulating cytokine levels in IBS have only been addressed in a small number of studies. Dinan et al. [25] reported increased plasma levels of IL-6 and IL-8 in IBS compared to controls, while plasma TNF-α and IL-10 levels did not differ significantly. Liebregts et al. [26], on the other hand, found an increase in both unstimulated and stimulated IL-6 production and a decrease in unstimulated TNF-α production by peripheral blood mononuclear cells in IBS. These increases were mainly attributable to diarrhoea-predominant (D-IBS) and PI-IBS groups [26].

Inflammation in Functional Dyspepsia

According to the Rome III consensus, FD is defined as the presence of early satiation, postprandial fullness or epigastric pain or burning, in the absence of any organic, systemic or metabolic disease that is likely to explain the symptoms [5]. Although different putative pathophysiological mechanisms such as delayed gastric emptying, impaired accommodation, visceral hypersensitivity, psychological factors and brain-gut axis dysfunction have been implicated, the underlying pathophysiology is poorly understood. In FD, evidence in favor of involvement of the immune system or a prior inflammatory event was only recently presented.

Histological Evidence

There is scarce evidence for the presence of inflammatory changes in functional dyspepsia. In 1989, Collins et al. [27] reported a significant increase in the number of polymorphonuclear and mononuclear cells in duodenal biopsies of non-ulcer dyspepsia patients as compared to healthy individuals, while no changes had been found previously in gastric biopsies [28]. More recently, the risk for functional dyspepsia appeared to be increased in the presence of an increased number of duodenal eosinophils in a population-based study from Scandinavia [29]. However, so far these changes have not been confirmed by others. In FD, however, IFN-γ production was significantly reduced, whereas there was no difference in IBS.

FD is a heterogeneous disorder. Although the exact pathophysiology remains largely unknown, different mechanisms, such as impaired gastric accommodation, increased duodenal acid exposure, delayed gastric emptying, visceral hypersensitivity and altered brain-gut axis interactions have been postulated to underlie the symptom pattern [6]. More recently, and analogous to findings in IBS, evidence for the role of a prior gastrointestinal inflammatory event has arisen. It has indeed been established that the prevalence of functional dyspepsia increased significantly following an outbreak of *Salmonella* gastroenteritis and remained elevated 1 year later [14]. There was also major overlap between new-onset FD and new-onset IBS after *Salmonella* gastroenteritis [14].

FD patients in whom symptoms start after an acute gastrointestinal infection report symptoms of early satiety and weight loss more frequently, while gastric sensorimotor function testing reveals an increased prevalence of impaired gastric accommodation [30]. Based on pharmacological studies in these patients, dysfunction of the gastric nitrergic innervation has been put forward as a putative pathophysiological mechanism. In contrast to healthy controls and unspecified onset FD, it was observed that sumatriptan, a 5-HT_{1D} receptor agonist thought to activate nitrergic neurons, did not evoke gastric relaxation in PI-FD patients [30]. On the other hand, the potent nitric oxide donor glyceryl trinitrate alleviated symptoms in FD [31].

While these studies indicate that acute gastrointestinal infections may trigger FD, through mechanisms that involve nitrergic nerve dysfunction, they do not establish whether this involves acute nerve damage with resolution of inflammation, or persisting gastrointestinal tract or systemic immune activation. The Leuven group studied duodenal biopsies in PI-FD and unspecified-onset FD [32]. These provided evidence for discrete changes in the cellular immunological response following a previous inflammatory event. Half of the PI-FD patients had focal aggregates of T cells, especially CD8+ cells, on duodenal biopsies, while these were not present in unspecified-onset FD. The total number of C8+ cells, however, remained unaffected. Additionally, the number of CD4+ cells in duodenal villi and around crypts was lower in PI-FD, whereas the number of CD68+ cells around crypts was higher in this group.

As biopsies were taken on average 10 months after symptom onset, it is unlikely that the observed changes represent the normal inflammatory process following the initial gastrointestinal infection [32].

Immune Function

Besides histological signs of immune activation in the duodenum, systemic cellular and humoral immune responses were also studied in PI-FD and unspecified-onset FD, and compared to IBS and non-cardiac chest pain [33].

Changes in the cytokine expression profile, such as elevated stimulated IL-5 and IL-13 production and decreased stimulated IFN-γ and IL-10 production, were comparable for FD and IBS, and some changes in immune function were more pronounced in the PI-FD group.

Pathophysiological Correlates of Immune Activation in Functional Gastrointestinal Disorders

The increased IL-6 production in IBS has been attributed to changes in the hypothalamic-pituitary-adrenal axis [25]. Anxiety and depression have also been implicated in altered immune function in several disease states, including FGID. In IBS, a significant correlation was found between anxiety scores and stimulated TNF production [26]. In FD, significant correlations between anxiety and depression scores and stimulated IL-5, IL-13 and TNF levels were found [33]. This is in line with extensive literature on the interaction between the immune system and psychological disorders [34, 35].

In IBS, signs of low-grade inflammation are associated with increased mucosal permeability [17]. In PI-FD, focal aggregates of CD8+ cells were associated with a higher prevalence of delayed gastric emptying and a tendency for more impairment of gastric accommodation [32]. No associations between immune activation and visceral hypersensitivity have been found [32].

Therapeutic Implications

It remains to be established whether immune activation in FGIDs is a target for therapy. A placebo-controlled trial with prednisolone in PI-IBS failed to demonstrate any beneficial effect [36]. Given the broad anti-inflammatory actions of steroids and their side effects, they are not very attractive to use in a benign condition like IBS.

Less potent, better targeted approaches to alter immune function may provide more attractive therapeutic applications in IBS. Older studies suggest a beneficial effect of mast cell stabilizers, like disodium cromoglycate, on symptoms in diarrhea-predominant IBS [37]. A recent preliminary study from Amsterdam, using ketotifen, confirmed the therapeutic potential of mast cell stabilization [38]. A recently published pilot study found that an 8-week course of mesalazine, widely used in inflammatory bowel disease, significantly reduced the number of colonic immune cells on biopsies in IBS patients, and this was associated with significantly improved general well-being and tendencies for improvement of abdominal pain [39].

Probiotics may provide another approach to improving immune activation in IBS. In a controlled study with *Bifidobacterium infantis* 35624 in IBS, this probiotic preparation not only improved composite and individual scores for abdominal pain/

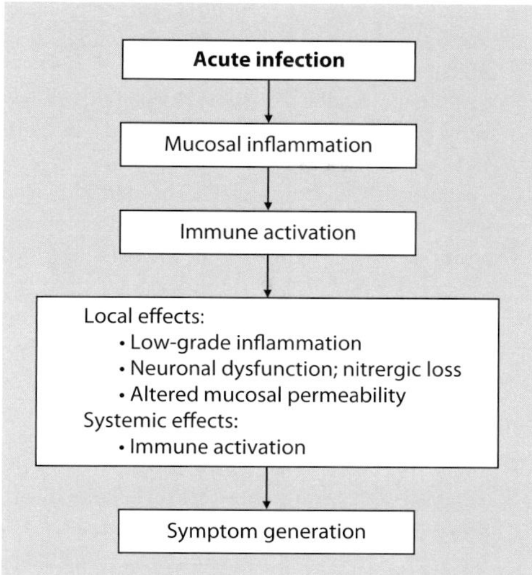

Fig. 1. Putative role of infection and inflammation in the pathogenesis of FGID.

discomfort, bloating/distention, and bowel movement difficulty, but also normalized IL-10/IL-12 ratios, which were abnormal at baseline [40].

Summary

There is now solid epidemiological evidence for acute infections as a trigger of functional gastrointestinal disorders (fig. 1). An increasing number of studies have shown signs of immune activation, both locally in the gastrointestinal tract and systemically in patients with (post-infectious) FGID. The underlying mechanisms and the exact role of immune activation in symptom generation remains to be determined. Further studies will be needed to establish whether targeting immune activation may lead to symptom improvement in FGIDs, and which would be the preferred targets in doing so.

References

1. Drossman DA: Functional GI disorders: what's in a name? Gastroenterology 2005;128:1771–1772.
2. Talley NJ: Functional gastrointestinal disorders as a public health problem. Neurogastroenterol Motil 2008;20(suppl 1):121–129.
3. Drossman DA: The functional gastrointestinal disorders and the Rome III process. Gastroenterology 2006;130:1377–1390.
4. Tack J, Talley NJ, Camilleri M, Holtmann G, Hu P, Malagelada JR, Stanghellini V: Functional gastroduodenal disorders. Gastroenterology 2006;130:1466–1479.
5. Longstreth GF, Thompson WG, Chey WD, Houghton LA, Mearin F, Spiller RC: Functional bowel disorders. Gastroenterology 2006;130:1480–1491.

6 Tack J, Bisschops R, Sarnelli G: Pathophysiology and treatment of functional dyspepsia. Gastroenterology 2004;127:1239–1255.
7 Camilleri M, Coulie B, Tack JF: Visceral hypersensitivity: facts, speculations, and challenges. Gut. 2001; 48:125–131.
8 Chaudhary NA, Truelove SC: The irritable colon syndrome: a study of the clinical features, predisposing causes, and prognosis in 130 cases. Q J Med 1962;31:307–322
9 McKendrick MW, Read NW: Irritable bowel syndrome: post *Salmonella* infection. J Infect 1994;29:1–3.
10 Neal KR, Hebden J, Spiller R: Prevalence of gastrointestinal symptoms six months after bacterial gastroenteritis and risk factors for development of the irritable bowel syndrome: postal survey of patients. BMJ 1997;314:779–782.
11 Parry SD, Stansfield R, Jelley D, Gregory W, Phillips E, Barton JR, Welfare MR: Does bacterial gastroenteritis predispose people to functional gastrointestinal disorders? A prospective, community-based, case-control study. Am J Gastroenterol 2003;98: 1970–1975.
12 Wang LH, Fang XC, Pan GZ: Bacillary dysentery as a causative factor of irritable bowel syndrome and its pathogenesis. Gut 2004;53:1096–1101.
13 Ji S, Park H, Lee D, Song YK, Choi JP, Lee SI: Postinfectious irritable bowel syndrome in patients with *Shigella* infection. J Gastroenterol Hepatol 2005;20: 381–386.
14 Mearin F, Perez-Oliveras M, Perello A, Vinyet J, Ibanez A, Coderch J, Perona M: Dyspepsia and irritable bowel syndrome after a *Salmonella* gastroenteritis outbreak: one-year follow-up cohort study. Gastroenterology 2005;129:98–104.
15 Marshall JK, Thabane M, Garg AX, Clark WF, Salvadori M, Collins SM: Incidence and epidemiology of irritable bowel syndrome after a large waterborne outbreak of bacterial dysentery. Gastroenterology 2006;131:445–450.
16 Gwee KA, Leong YL, Graham C, McKendrick MW, Collins SM, Walters SJ, Underwood JE, Read NW: The role of psychological and biological factors in postinfective gut dysfunction. Gut 1999;44:400–406.
17 Spiller RC, Jenkins D, Thornley JP, Hebden JM, Wright T, Skinner M, Neal KR: Increased rectal mucosal enteroendocrine cells, T lymphocytes, and increased gut permeability following acute *Campylobacter* enteritis and in post-dysenteric irritable bowel syndrome. Gut 2000;47:804–811.
18 Dunlop SP, Jenkins D, Neal KR, Spiller RC: Relative importance of enterochromaffin cell hyperplasia, anxiety, and depression in postinfectious IBS. Gastroenterology 2003;125:1651–1659.
19 Ohman L, Isaksson S, Lundgren A, Simrén M, Sjövall H: A controlled study of colonic immune activity and beta7+ blood T lymphocytes in patients with irritable bowel syndrome. Clin Gastroenterol Hepatol 2005;3:980–986.
20 Barbara G, Stanghellini V, De Giorgio R, Cremon C, Cottrell GS, Santini D, Pasquinelli G, Morselli-Labate AM, Grady EF, Bunnett NW, Collins SM, Corinaldesi R: Activated mast cells in proximity to colonic nerves correlate with abdominal pain in irritable bowel syndrome. Gastroenterology 2004; 126:693–702.
21 Kristjánsson G, Venge P, Wanders A, Lööf L, Hällgren R: Clinical and subclinical intestinal inflammation assessed by the mucosal patch technique: studies of mucosal neutrophil and eosinophil activation in inflammatory bowel diseases and irritable bowel syndrome. Gut 2004;53:1806–1812.
22 Gwee KA, Collins SM, Read NW, Rajnakova A, Deng Y, Graham JC, McKendrick MW, Moochhala SM: Increased rectal mucosal expression of interleukin 1beta in recently acquired post-infectious irritable bowel syndrome. Gut 2003;52:523–526.
23 Gonsalkorale WM, Perrey C, Pravica V, Whorwell PJ, Hutchinson IV: Interleukin 10 genotypes in irritable bowel syndrome: evidence for an inflammatory component? Gut 2003;52:91–93.
24 van der Veek PP, van den Berg M, de Kroon YE, Verspaget HW, Masclee AA: Role of tumor necrosis factor-alpha and interleukin-10 gene polymorphisms in irritable bowel syndrome. Am J Gastroenterol 2005;100:2510–2516.
25 Dinan TG, Quigley EM, Ahmed SM, Scully P, O'Brien S, O'Mahony L, O'Mahony S, Shanahan F, Keeling PW: Hypothalamic-pituitary-gut axis dysregulation in irritable bowel syndrome: plasma cytokines as a potential biomarker? Gastroenterology 2006;130:304–311.
26 Liebregts T, Adam, B, Bredack C, Röth A, Heinzel S, Lester S, Downie-Doyle S, Smith E, Drew P, Talley NJ, Holtmann G: Immune activation in patients with irritable bowel syndrome. Gastroenterology 2007;132:913–920.
27 Collins JS, Hamilton PW, Watt PC, Sloan JM, Love AH: Superficial gastritis and *Campylobacter pylori* in dyspeptic patients: a quantitative study using computer-linked image analysis. J Pathol 1989;158: 303–310.
28 Collins JS, Hamilton PW, Watt PC, Sloan JM, Love AH: Quantitative histological study of mucosal inflammatory cell densities in endoscopic duodenal biopsy specimens from dyspeptic patients using computer linked image analysis. Gut 1990;31:858–861.

29 Talley NJ, Walker MM, Aro P, Ronkainen J, Storskrubb T, Hindley LA, Harmsen WS, Zinsmeister AR, Agréus L: Non-ulcer dyspepsia and duodenal eosinophilia: an adult endoscopic population-based case-control study. Clin Gastroenterol Hepatol. 2007;5:1175–1183.

30 Tack J, Demedts I, Dehondt G, Caenepeel P, Fischler B, Zandecki M, Janssens J: Clinical and pathophysiological characteristics of acute-onset functional dyspepsia. Gastroenterology 2002;122:1738–1747.

31 Gilja OH, Hausken T, Wilhelmsen I, Berstad A: Impaired accommodation of proximal stomach to a meal in functional dyspepsia. Dig Dis Sci 1996;41:689–696.

32 Kindt S, Tertychnyy A, de Hertogh G, Geboes K, Tack J: Intestinal immune activation in presumed post-infectious functional dyspepsia. Neurogastroenterol Motil 2009;21:832–e56.

33 Kindt S, Van Oudenhove L, Broekaert D, Kasran A, Ceuppens JL, Bossuyt X, Fischler B, Tack J: Immune dysfunction in patients with functional gastrointestinal disorders. Neurogastroenterol Motil 2009;21:389–398.

34 Strouse TB: The relationship between cytokines and pain/depression: a review and current status. Curr Pain Headache Rep 2007;11:98–103.

35 Jones JF: An extended concept of altered self: chronic fatigue and post-infection syndromes. Psychoneuroendocrinology 2008;33:119–129.

36 Dunlop SP, Jenkins D, Neal KR, Naesdal J, Borgaonker M, Collins SM, Spiller RC: Randomized, double-blind, placebo-controlled trial of prednisolone in post-infectious irritable bowel syndrome. Aliment Pharmacol Ther 2003;18:77–84.

37 Stefanini GF, Prati E, Albini MC, Piccinini G, Capelli S, Castelli E, Mazzetti M, Gasbarrini G: Oral disodium cromoglycate treatment on irritable bowel syndrome: an open study on 101 subjects with diarrheic type. Am J Gastroenterol 1992;87:55–57.

38 Klooker TK, Koopman KEM, vd Heide S, vd Wijngaard RM, Boeckxstaens GE: Treatment with the mast cell stabilizer ketotifen decreases visceral hypersensitivity and improves intestinal symptoms in IBS patients. Gut 2008;57(suppl 2):A86 (abstract).

39 Corinaldesi R, Stanghellini V, Cremon C, Gargano L, Cogliandro RF, De Giorgio R, Bartesaghi G, Canovi B, Barbara G: Effect of mesalazine on mucosal immune biomarkers in irritable bowel syndrome: a randomized controlled proof-of-concept study. Aliment Pharmacol Ther 2009;30:245–252.

40 O'Mahony L, McCarthy J, Kelly P, Hurley G, Luo F, Chen K, O'Sullivan GC, Kiely B, Collins JK, Shanahan F, Quigley EM: *Lactobacillus* and *Bifidobacterium* in irritable bowel syndrome: symptom responses and relationship to cytokine profiles. Gastroenterology 2005;128:541–551.

Jan Tack, MD, PhD
Department of Pathophysiology, Gastroenterology Section
University of Leuven
Herestraat 49, BE–3000 Leuven (Belgium)
Tel. +32 16 344 225, Fax +32 16 344 419, E-Mail jan.tack@med.kuleuven.be

Mayerle J, Tilg H (eds): Clinical Update on Inflammatory Disorders of the Gastrointestinal Tract.
Front Gastrointest Res. Basel, Karger, 2010, vol 26, pp 126–134

Probiotics in GI Diseases

P. Gionchetti · F. Rizzello · R. Tambasco · R. Brugnera · G. Straforini · S. Nobile · G. Liguori · G. Spuri Fornarini · M. Campieri

Department of Clinical Medicine, University of Bologna, Bologna, Italy

Abstract

Probiotics are living organisms, which upon ingestion in certain numbers, exert health benefits beyond inherent basic nutrition. Probiotics have been studied in a variety of GI diseases, and are an appealing concept given their favourable safety profiles. Current data on the use of probiotic therapy in GI diseases are reviewed. The rationale for using probiotics in IBD is based on convincing evidence that implicates intestinal bacteria in the pathogenesis of the disease. VSL#3, a highly concentrated cocktail of probiotics, has been shown to be effective in the prevention of pouchitis onset and relapses. Results from the use of probiotics in ulcerative colitis have been promising, while results in Crohn's disease are not yet clear. Placebo-controlled trials indicated that lactobacilli have a suppressive effect on *Helicobacter pylori* infection. Although some studies reported improvement in *H. pylori* eradication, others failed to confirm this. The clear delineation of a post-infective variety of irritable bowel syndrome (IBS), suggests a role for a dysfunctional relationship between the indigenous flora and the host in IBS and, accordingly, provides a clear rationale for the use of probiotics in this disorder. Probiotics may have a role in alleviating some of the symptoms of IBS, but results from controlled trials are controversial. Future studies are needed, in particular larger studies of longer duration with greater methodological rigor. Controlled trials support the use of *Lactobacillus rhamnosus GG* and *Saccharomyces boulardii* for the prevention of antibiotic-associated diarrhoea, and have demonstrated the effectiveness of *S. boulardii* as adjunctive therapy in *Clostridium difficile* disease.

Probiotics

The potential benefit of probiotics in health maintenance and disease prevention has long been acknowledged. At the turn of the last century, the Russian Nobel Prize winner Elie Metchnikoff suggested that high concentrations of lactobacilli in the intestinal flora were important for health and longevity in humans. Probiotics are defined as living organisms, which upon ingestion in certain numbers, exert health benefits beyond inherent basic nutrition.

The bacteria most commonly associated with probiotic activity are lactobacilli, bifidobacteria, and streptococci, but other non-pathogenic bacteria (e.g. some strains of *Escherichia coli*) and non-bacterial organisms (e.g. the yeast *Saccharomyces boulardii*) have been used.

Several mechanisms have been proposed to account for the action of probiotics. These include antagonistic activity against pathogenic bacteria, either by inhibition of adherence and translocation, or by production of antibacterial substances such as antimicrobial peptides (bacteriocins) and hydrogen peroxide. Probiotics also stimulate mucosal defence, both at the level of immune and epithelial function, by increasing production of secretory immunoglobulin A, blocking pro-inflammatory cytokines, enhancing levels of anti-inflammatory cytokines, stimulating expression of intestinal mucins and improving gut permeability.

Additionally, they are able to produce nutrients of special importance to the intestine, such as short-chain fatty acids and vitamins.

Inflammatory Bowel Disease

The rationale for using probiotics in IBD is based on convincing evidence that implicates intestinal bacteria in the pathogenesis of the disease [1]. The distal ileum and the colon are the areas with the highest bacterial concentrations and represent the sites of inflammation in IBD. Similarly, pouchitis, the non-specific inflammation of the ileal reservoir after ileo-anal anastomosis appears to be associated with bacterial overgrowth and dysbiosis. Enteric bacteria and their products have been found within the inflamed mucosa of patients with Crohn's disease (CD).

There is evidence of a loss of immunological tolerance to commensal bacteria in patients with IBD. Patients with CD consistently respond to diversion of faecal stream, with immediate recurrence of inflammation after restoration of intestinal continuity or infusion of luminal content into the bypassed ileum. Furthermore, pouchitis does not occur prior to closure of the ileostomy.

The composition of the enteric flora is altered in patients with IBD. Increased numbers of aggressive bacteria, such as *Bacteroides*, adherent/invasive *E. coli*, and enterococci, and decreased numbers of protective lactobacilli and bifidobacteria have been observed. The most compelling evidence that intestinal bacteria play a role in IBD is derived from animal models. Despite great diversity in genetic defects and immunopathology, a consistent feature of many transgenic and knockout mutant murine models of colitis is that the presence of normal enteric flora is required for full expression of inflammation.

All of these observations suggest that IBD may be prevented or treated by the manipulation of intestinal microflora, and increasing evidence supports a therapeutic role for probiotics and prebiotics in IBD.

Ulcerative Colitis

Three trials have evaluated the efficacy of the probiotic preparation *E. coli Nissle* 1917 in the treatment of IBD. In the 3 controlled studies, *E. coli Nissle* 1917 was shown to be as efficacious as mesalazine as a maintenance treatment for ulcerative colitis (UC) [2–4].

In a small randomized, controlled trial study 21 patients with UC in remission were treated with bifidobacteria-fermented milk or placebo for 12 months. At the end of 12 months clinical remission was observed 8 of 11 (73%) taking fermented milk and in 1 of 10 control subjects (10%) [5]. We have investigated the use of VSL#3 in the treatment of UC. This product contains cells of 4 strains of lactobacilli *(L. casei, L. plantarum, L. acidophilus, L. delbrueckii bulgaricus)*, 3 strains of bifidobacteria *(B. longum, B. breve, B. infantis)*, and 1 strain of *Streptococcus salivarius thermophilus*. Each packet of VSL#3 contains 450 billion viable lyophilized bacteria. A pilot study was performed using VSL#3 as a maintenance treatment in UC patients in remission who were either allergic or intolerant to sulphasalazine and mesalazine. Patients (n = 20) received VSL#3 1,800 billion bacteria/day for 12 months and were assessed clinically and endoscopically at baseline, at 6 and 12 months, or if relapse occurred. Stool culture was also performed at different intervals. Faecal concentrations of lactobacilli, bifidobacteria, and *S. thermophilus* were significantly increased by VSL#3. This increase was evident from day 20 and persisted throughout the treatment period. Levels returned to baseline within 15 days of treatment cessation. In total, 15 of the 20 patients (75%) remained in remission during the study [6]. More recently, an open-label study evaluated the efficacy of high-dose VSL#3 (3,600 billion bacteria) for 6 weeks in 34 patients with active mild to moderate UC, who had failed mesalazine therapy. Intent to treat analysis showed remission in 53% and a positive response in a further 24% [7]. Guslandi et al. [8] studied 25 patients with active mild-moderate flare of UC while on maintenance treatment with mesalazine 3 g per day, in a 4-week, open-label, uncontrolled study. Patients received the yeast *S. boulardii* 750 mg per day. A remission was observed in 17 of 25 patients (68%).

Pouchitis

Pouchitis, a non-specific, idiopathic inflammation of the ileal reservoir, has become the most frequent long-term complication following pouch surgery for ulcerative colitis. The aetiology of pouchitis is still unknown, and is likely to be multifactorial; however, the immediate response to antibiotic treatment suggests a pathogenic role for the microflora, and recently pouchitis was associated with a decreased ratio of anaerobic to aerobic bacteria, reduced faecal concentrations of lactobacilli and bifidobacteria, and an increase of luminal pH. Treatment of pouchitis is largely empiric and only a few small placebo-controlled trials have been conducted. Antibiotics are the mainstay of treatment, and metronidazole and ciprofloxacin are the common initial therapeutic approach and most patients have a dramatic response within few days.

We have carried out a double-blind study to compare the efficacy of VSL#3 with placebo in the maintenance treatment of chronic pouchitis [9]. Patients (n = 40) who were in clinical and endoscopic remission after 1 month of combined antibiotic treatment (2 g/day of rifaximin plus 1 g/day of ciprofloxacin) were randomized to receive either VSL#3 1,800 billion bacteria/day or placebo for 9 months. Patients were assessed clinically every month, and assessed endoscopically and histologically at entry and every 2 months thereafter. Stool culture was performed before and after antibiotic treatment, and monthly during maintenance treatment. Relapse was defined as an increase of at least 2 points in the clinical section of the Pouchitis Disease Activity Index (PDAI) and was confirmed endoscopically and histologically. All 20 patients treated with placebo relapsed during the follow-up period. In contrast, 17 of the 20 (85%) patients treated with VSL#3 were still in remission after 9 months. Interestingly, all 17 of these patients relapsed within 4 months of suspension of the active treatment. Faecal concentrations of lactobacilli, bifidobacteria and *S. thermophilus* were significantly increased within 1 month of treatment initiation and remained stable throughout the study. This increase did not affect the concentration of the other bacterial groups, which suggests that the effect was not mediated by suppression of endogenous luminal bacteria. A subsequent double-blind, placebo-controlled study on the effectiveness of VSL#3 in the maintenance of antibiotic-induced remission in patients with refractory or recurrent pouchitis reported similar results [10]. After 1 year of treatment, 85% of those in the VSL#3 group were in remission, versus only 6% of those in the placebo group. In contrast to the positive results reported in trials on VSL#3, a double-blind, placebo-controlled trial on *Lactobacillusrhamnosus* strain GG *(Lactobacillus GG)* in patients with a previous history of pouchitis revealed that this probiotic was not effective in preventing relapses [11].

A double-blind, placebo-controlled trial evaluated the efficacy of VSL#3 in the prevention of pouchitis onset in patients following ileal-pouch anal anastomosis for UC [12]. Within 1 week after ileostomy closure, 40 patients were randomized to receive either VSL#3 (3 g/day, equivalent to 900 billion bacteria) or placebo for 12 months. Patients were assessed clinically, endoscopically and histologically at 1, 3, 6, 9 and 12 months according to PDAI score. During the first year after ileostomy closure, patients treated with VSL#3 had a significantly lower incidence of acute pouchitis compared with those treated with placebo (10 vs. 40%, $p < 0.05$). Moreover, the IBD questionnaire score was significantly improved only in the group treated with VSL#3 and among those who did not develop pouchitis, the median stool frequency was significantly lower in the VSL#3 group.

Results with probiotics in CD are conflicting. In a small pilot study, *E. coli Nissle* 1917 was compared with placebo in the maintenance of steroid-induced remission of colonic CD [13]. Twelve patients were treated with *E. coli Nissle* 1917 and 11 were treated with placebo. At the end of the 12-week treatment period, relapse rates were 33% in the *E. coli* group and 63% in the placebo group; unfortunately, due the very small number of patients treated, this difference did not reach statistical significance. In a small, comparative, 6-month, open-label study, combination therapy with the yeast *S. boulardii* (1 g/day) plus mesalamine (2 g/day) was significantly superior to monotherapy with mesalamine (3 g/day) in the maintenance of remission [14]. In a 1-year, double-blind, placebo-controlled trial, *Lactobacillus GG* was not effective in the prevention of post-operative recurrence [15]. Similarly in a double-blind trial *Lactobacillus GG* was shown to be not superior to placebo in prolonging remission in children with CD when given as an adjunct to standard therapy [16].

Two randomized double-blind, placebo-controlled studies showed *Lactobacillus johnsonii* LA1 (4×10^9 cfu/day) was not superior to placebo to prevent endoscopic recurrence of CD [17, 18].

We performed a single-blind study to compare a sequential antibiotic-probiotic treatment with mesalazine in the prevention of post-operative recurrence of CD. Within 1 week after curative surgery, 40 patients were randomized to receive either high-dose rifaximin (an unabsorbable wide-spectrum antibiotic) for 3 months followed by VSL#3 (6 g/day) for 9 months, or mesalazine (4 g/day) for 12 months. Patients were assessed clinically and endoscopically at 3 and 12 months. Compared with placebo, the combined antibiotic-probiotic treatment was associated with a significantly lower incidence of severe endoscopic recurrence, both at 3 months (10 vs. 40%, $p < 0.01$) and 12 months (20 vs. 40%, $p < 0.01$) [19].

H. pylori Infection

Antibiotic-associated gastrointestinal side-effects such as diarrhea, nausea, vomiting, bloating and abdominal pain may represent a serious drawback of anti-H. pylori therapies, even though they are mild in most cases, usually resulting in discontinue therapy. These manifestations have been related to quantitative and qualitative changes in the intestinal microflora due to unabsorbed or secreted antibiotics in the intestinal content, with a resulting reduction in normal saprophytic flora and overgrowth and the persistence of potentially pathogenic antibiotic-resistant indigenous strains. Several studies have reported that certain probiotic bacteria, such as *Lactobacillus* species, exhibit inhibitory activity against *H. pylori* in vitro and in vivo.

A recent meta-analysis has suggested that probiotic supplementation could be effective in increasing eradication rates of anti-*H. pylori* therapy; furthermore, probiotics showed a positive impact on *H. pylori* therapy-related side effects [20].

Irritable Bowel Syndrome

A rationale for the use of probiotics for a number of functional gastrointestinal symptoms and syndromes can be developed, and an experimental basis for their use continues to emerge, but data from well-conducted clinical trials of probiotics in this area remain scarce. Irritable bowel syndrome (IBS), a common disorder in which available therapies have limited efficacy, has attracted the most attention. Recent revelations regarding the potential pathogenic roles of the enteric flora and immune activation have led to a re-awakening of interest in bacterio-therapy for this common and challenging disorder.

Recent systematic reviews and meta-analysis have been carried-out to evaluate the efficacy, safety and tolerability of probiotics in the treatment of IBS.

In the first study, 20 trials with 23 probiotic treatment arms and a total of 1,404 subjects met inclusion criteria. Probiotic use was associated with improvement in global IBS symptoms compared to placebo [pooled relative risk (RR) 0.77, 995% CI 0.62–0.94]. Probiotics were also associated with less abdominal pain compared to placebo (pooled RR = 0.78, 95% CI 0.69–0.88). The authors concluded that, although the analyses suggested that probiotic use may be associated with improvement in IBS symptoms compared to placebo, the results should be interpreted with caution, given the methodological limitations of contributing studies [21].

Included in the second study were only parallel group randomized controlled trials (RCTs) with at least 1 week of therapy comparing probiotics with placebo or no treatment in adults with IBS according to any acceptable definition. Studies had to provide abdominal pain or global IBS symptom improvement as an outcome. Nineteen trials were identified, and the trial quality was generally good. There were 10 RCTs involving 918 patients providing dichotomous data. Probiotics were statistically significantly better than placebo (RR of IBS not improving = 0.71, 95% CI 0.57–0.88) with NNT = 4 (95% CI = 3–12.5). Authors concluded that probiotics appear to be efficacious in IBS but the magnitude of benefit and the most effective species and strain are uncertain [22].

In the third study, a total of 16 RCTs met selection criteria. Of those, 11 studies showed suboptimal study design with inadequate blinding, inadequate trial length, inadequate sample size, and/or lack of intention-to-treat analysis. None of the studies provided quantifiable data about both tolerability and adverse events. *Bifidobacterium infantis* 35624 showed significant improvement in the composite score for abdominal pain/discomfort, bloating/distention and/or bowel movement difficulty compared with placebo ($p < 0.05$) in 2 appropriately designed studies. No other probiotic showed significant improvement in IBS symptoms in an appropriately designed study [23]. In the most recently published systematic review, 14 randomized placebo controlled trials were identified. Combined data suggested a modest improvement in overall symptoms after several weeks of treatment [24]

Diverticular Disease

Treatment of diverticular disease is aimed at the relief of symptoms (such as abdominal pain, discomfort, bloating) and to prevent major complications. Some observations suggest a possible role of gut microflora in determining symptoms related to diverticular disease. Only 1 controlled trial has been carried out with probiotics in diverticular disease. The non-pathogenic *E. coli Nissle* (Mutaflor capsules, 2.5×10^{10} viable bacteria/capsule) significantly improved the symptoms in uncomplicated diverticular disease [25]. Large double-blind, placebo controlled trials are now recommended.

Antibiotic-Associated and *C. difficile*-Induced Diarrhoea

A recent meta-analysis has evaluated the efficacy of probiotics in the prevention of antibiotic-associated diarrhoea (AAD) and in the treatment or prevention of *Clostridium difficile* disease (CCD). Thirthy-one studies met the selection criteria. From 25 randomized controlled trials, probiotics significantly reduced the relative risk of AAD (RR = 0.43, 95% CI 0.31–0.59, $p < 0.001$). From six randomized trials, probiotics had significant efficacy for CCD (RR = 0.59, 95% CI 0.41–0.85, $p = 0.005$). Using meta-analyses, 3 types of probiotics (*S. boulardii, Lactobacillus GG* and probiotic mixture) significantly reduced the development of AAD; only *S. boulardii* was effective for CCD [26].

Conclusions

Probiotics may provide a simple and attractive way of preventing or treating IBD, and patients find the probiotic concept appealing because it is safe, nontoxic and natural. VSL#3, a highly concentrated cocktail of probiotics has been shown to be effective in the prevention of pouchitis onset and relapses. Results on the use of probiotics in UC are promising, both in terms of the prevention of relapses and the treatment of mild-to-moderate attacks. Results in CD are not yet clear because of conflicting data and the limited number of well-performed studies.

Probiotic supplementation could be effective in increasing eradication rates of anti-*H. pylori* therapy and have a positive impact on *H. pylori* therapy-related side effects.

Probiotics may have a role in alleviating some of the symptoms of IBS, a condition for which currently evidence of efficacy of drug therapies is weak. However, as IBS is a condition that is chronic and usually intermittent, longer term trials are recommended. Such research should focus on the type, optimal dose of probiotics and the subgroups of patients who are likely to benefit the most.

Future studies are needed, in particular larger studies of longer duration with greater methodological rigor. In addition, more data are needed regarding which specific strains and doses are most likely to be effective. The use of probiotics for IBS warrants further study, particularly given the chronic nature of this condition, its major impact on patients' quality of life, and the dearth of other effective treatments. Future probiotic studies should follow Rome III recommendations for appropriate design of RCTs.

Data on the efficacy of probiotics in diverticular disease are scarce.

Probiotics may have a role in preventing and treating antibiotic-associated diarrhoea and *C. difficile* disease.

It is important to select a well-characterized probiotic preparation, in view of the fact that the viability and survival of bacteria in many of the currently available preparations are unproven. It should noted that the beneficial effect of one probiotic preparation does not imply efficacy of other preparations containing different bacterial strains, because each individual probiotic strain has unique biological properties.

References

1 Campieri M, Gionchetti P: Probiotics in inflammatory bowel disease: new insight to pathogenesis or a possible therapeutic alternative? Gastroenterology 1999;116:1246–1249.

2 Kruis W, Schuts E, Fric P, et al: Double-blind comparison of an oral *Escherichia coli* preparation and mesalazine in maintaining remission of ulcerative colitis. Aliment Pharmacol Ther 1997;11:853–858.

3 Rembacken BJ, Snelling AM, Hawkey P, et al: Nonpathogenic *Escherichia coli* vs. mesalazine for the treatment of ulcerative colitis: a randomised trial. Lancet 1999;354:635–639.

4 Kruis W, Fric P, Pokrotnieks J, et al: Maintaining remission of ulcerative colitis with *Escherichia Coli Nissle* 1917 is as effective as with standard mesalazine. Gut 2004;53:1673–1623.

5 Ishikawa H, Akedo I, Umesaki Y, et al: Randomized, controlled trial of the effect of bifidobacteria-fermented milk on ulcerative colitis. J Am Coll Nutr 2003;22:56–63.

6 Venturi A, Gionchetti P, Rizzello F, et al: Impact on the faecal flora composition of a new probiotic preparation: preliminary data on maintenance treatment of patients with ulcerative colitis (UC) intolerant or allergic to 5-aminosalicylic acid (5 ASA). Aliment Pharmacol Ther 1999;13:1103–1108.

7 Bibiloni R, Fedorak RN, Tannock GW, Madsen KL, Gionchetti P, Campieri M, De Simone C, Sartor RB: VSL#3 probiotic mixture induces remission in patients with active ulcerative colitis. Am J Gastroenterol 2005;100:1539–1546.

8 Guslandi M, Giollo P, Testoni PA: A pilot trial of *Saccharomyces boulardii* in ulcerative colitis. Eur J Gastroenterol Hepatol 2003;15:697–698.

9 Gionchetti P, Rizzello F, Venturi A, et al: Oral bacteriotherapy as maintenance treatment in patients with chronic pouchitis: a double-blind, placebo-controlled trial. Gastroenterology 2000;119:305–309.

10 Mimura T, Rizzello F, Helwig U, et al: Once daily high dose probiotic therapy for maintaining remission in recurrent or refractory pouchitis. Gut 2004; 53:108–114.

11 Kuisma J, Mentula S, Kahri A, et al: Effect of *Lactobacillus rhamnosus GG* on ileal pouch inflammation and microbial flora. Aliment Pharmacol Ther 2003;17:509–515.

12 Gionchetti P, Rizzello F, Helvig U, Venturi A, Lammers KM, Brigidi P, et al: Prophylaxis of pouchitis onset with probiotic therapy: a double-blind placebo controlled trial. Gastroenterology 2003;124:1202–1209.

13 Malchow HA: Crohn's disease and *Escherichia coli*: a new approach in therapy to maintain remission of colonic Crohn's disease? J Clin Gastroenterol 1997; 25:653–658.

14 Guslandi M, Mezzi G, Sorghi M, et al: *Saccharomyces boulardii* in maintenance treatment of Crohn's disease. Dig Dis Sci 2000;45:1462–1464.

15 Prantera C, Scribano ML, Falasco G, et al: Ineffectiveness of probiotics in preventing recurrence after curative resection for Crohn's disease: a randomised controlled trial with *Lactobacillus GG*. Gut 2002;51;405–409.

16 Marteau P, Lemann M, Seksik P, Laharie D, Colombel JF, Bouhnik Y, Cadiot G, Soule JC, Boureille A, Metman E, Lerebours E, Carbonnel F, Dupas JL, Vevrac M, Coffin B, Moreau J, Abitbol V, Blum-Sperisen S, Mary JY: Ineffectiveness of *Lactobacillus johnsonii* LA1 for prophilaxis of postoperative recurrence in Crohn's disease: a randomized, double-blind, placebo-controlled GETAID trial. Gut 2005.

17 Van Gossum A, Dewit O, Louis E, de Hertogh G, Baert F, Fontaine F, DeVos M, Enslen M, Paintin M, Franchimont D: Multicenter randomized-controlled clinical trial of probiotics (*Lactobacillus johnsonii*, LA1) on early endoscopic recurrence of Crohn's disease after ileo-caecal resection. Inflamm Bowel Dis 2007;13:135–142.

18 Bousvaros A, Guandalini S, Baldassano RN, Botelho C, Evans J, Ferry GD, Golden B, Hartigan L, Kugathasan S, Levy J, Murray KF, Oliva-Hemker M, Rosh JR, Tolia V, Zholudev A, Vanderhoof JA, Hibberd PL: A randomized, double-blind trial of Lactobacillus GG versus placebo in addition to standard maintenance therapy for children with Crohn's disease. Inflamm Bowel Dis 2005;11:833–839.

19 Campieri M, Rizzello F, Venturi A, et al: Combination of antibiotic and probiotic treatment is efficacious in prophylaxis of post-operative recurrence of Crohn's disease: a randomised controlled study vs. mesalazine. Gastroenterology 2000;118: A781.

20 Tong JL, Ran ZH, Shen J, Zhang CX, Xiao SD: Meta-analysis: the effect of probiotics supplementation on eradication rates and adverse events during *Helicobacter pylori* eradication therapy. Aliment Pharmacol Ther 2007;25:155–168.

21 McFarland LV, Dublin S: Meta-analysis of probiotics for the treatment of irritable bowel syndrome. World J Gastro 2008;14:2650–2661.

22 Moayyedi P, Ford AC, Talley NJ, Cremonini F, Foxx-Orenstein A, Brandt L, Quigley E: The efficacy of probiotics in the therapy of irritable bowel syndrome: a systematic review. Gut 2008, Epub ahead of print.

23 Hoveyda N, Heneghan C, Mahtani KR, Perera R, Roberts N, Glasziou P: A systematic review and meta-analysis: probiotics in the treatment of irritable bowel syndrome. BMC Gastroenterol 2009;16:9–15.

24 Brenner DM, Moeller MJ, Chey WD, Schoenfeld PS: The utility of probiotics in the treatment of irritable bowel syndrome: a systematic review. Am J Gastroenterol 2009;104:1033–1049.

25 Fric P, Zavoral M: The effect of non-pathogenic *Escherichia coli* in symptomatic uncomplicated diverticular disease of the colon. Eur J Gastroenterol Hepatol 2003;15:313–315.

26 McFarland LV: Meta-analysis of probiotics for the prevention of antibiotic-associated diarrhoea and the treatment of *Clostridium difficile* disease. Am J Gastroenterol 2006;101:812–822.

Dr. Paolo Gionchetti, MD
IBD Unit, Department of Clinical Medicine
S. Orsola-Malpighi Hospital, University of Bologna
via Massarenti 9, IT–40138 Bologna (Italy)
Tel. +39 051 636 4122 ext. 4102, Fax +39 051 636 4102 392 538, E-Mail paolo.gionchetti@unibo.it

Mayerle J, Tilg H (eds): Clinical Update on Inflammatory Disorders of the Gastrointestinal Tract.
Front Gastrointest Res. Basel, Karger, 2010, vol 26, pp 135–145

Microscopic Colitis

Darrell S. Pardi[a] · Stephan Miehlke[b]

[a]Inflammatory Bowel Disease Clinic, Division of Gastroenterology and Hepatology, Mayo Clinic College of Medicine, Rochester, Minn., USA, and [b]Medical Department I, Technical University Hospital, Dresden, Germany

Abstract

Microscopic colitis is a common cause of chronic diarrhoea, particularly in the elderly. The colonic mucosa appears normal or nearly normal at endoscopy, and the diagnosis is made on histologic grounds in the appropriate clinical setting. There are 2 subtypes, collagenous and lymphocytic colitis, that are clinically and histologically similar, but which can be differentiated by the presence or absence of a thickened subepithelial collagen band. The entity of drug-induced microscopic colitis and/or concomitant celiac sprue needs to be considered when evaluating these patients. There are few controlled treatment trials in microscopic colitis, with budesonide being the best studied drug. Despite this fact, systematic approach to therapy often leads to satisfactory control of symptoms.

Background

The term 'microscopic colitis' was originally used to describe patients with chronic diarrhoea and inflammation on colon biopsies but normal findings on sigmoidoscopy and barium enema [1]. Collagenous colitis is a related condition with similar clinical and histologic features, but with the additional finding of a thickened subepithelial collagen band [2]. It is unclear whether these 2 conditions are different diseases or just part of a spectrum of the same disease. Indeed, reviews of some of the early cases of 'microscopic colitis' showed that many actually had collagenous colitis [3, 4]. Since the colon in collagenous colitis is grossly normal, it is also considered a form of 'microscopic' colitis. Thus, microscopic colitis is used as an umbrella term with 2 subtypes: collagenous colitis, in which the subepithelial collagen band is thickened, and lymphocytic colitis, in which the collagen band is normal [4].

Epidemiology

Microscopic colitis has been found in 4–13% of patients investigated for chronic diarrhoea [5–8]. In Europe and North America, the reported incidence of collagenous colitis is 0.6–5.2/100,000 and for lymphocytic colitis it is 3.7–5.5/100,000 [7–11]. In some of these studies, there was a significant increase in incidence over time (for example, from 0.8/100,000 in 1985–1989 to 19.1/100,000 in 1998–2001 in one study from North America [8]).

A female predominance has been described, particularly for collagenous colitis, with female to male ratios as high as 20:1 [7–14]. The gender difference for lymphocytic colitis is less striking than for collagenous colitis in some studies [8] but not others [11]. Microscopic colitis is increasingly common with age, with the diagnosis most typically made in the sixth or seventh decade [7–14], although a wide age range has been reported, including paediatric cases [15, 16]. There are reports of familial occurrence [17, 18], including in twins, although familial clusters are uncommon.

No association between microscopic colitis and colon cancer has been discovered [19, 20], but long-term studies are needed to further explore this possibility. Cases of lung cancer have been reported in collagenous colitis [12, 19], perhaps related to cigarette smoking, which is more common in collagenous than lymphocytic colitis or controls [21, 22].

Clinical Features

Microscopic colitis is characterized by chronic or intermittent watery diarrhoea, ranging from relatively mild and self-limited to severe, with dehydration and other metabolic consequences. Quality of life is affected in proportion to the amount of diarrhoea and faecal incontinence [23, 24]. Many patients have abdominal pain or weight loss. The weight loss is typically mild but can be significant in some cases [12, 25]. It is important to recognize that symptoms of microscopic colitis are non-specific. In fact, many patients with microscopic colitis meet the symptom-based criteria for irritable bowel syndrome [26, 27]. Therefore, these criteria cannot adequately distinguish these 2 diarrhoeal illnesses, which requires colonic mucosal biopsies. Faecal leukocytes may be present [25, 26], but steatorrhoea, fever or haematochezia should suggest an alternate diagnosis.

Arthralgias and various autoimmune conditions (e.g. thyroid dysfunction, diabetes, psoriasis) are often seen in patients with microscopic colitis [12, 14, 25]. In addition, an elevated erythrocyte sedimentation rate and a positive antinuclear antibody or other autoimmune markers [25, 28] have been reported.

Of interest is the association between microscopic colitis and coeliac sprue. Among patients with coeliac sprue, approximately one third have histologic changes in the colonic mucosa consistent with microscopic colitis [29, 30]. Thus, microscopic colitis is common in patients with coeliac disease, and this diagnosis should be considered in patients with coeliac sprue who have continued or recurrent diarrhoea despite a strict gluten-free diet [31].

On the other hand, small bowel sprue-like changes in patients with microscopic colitis are seen in 2–9% of cases in the largest series [12, 14, 25]. Anti-endomysial and anti-tissue transglutaminase antibodies are found infrequently in patients with microscopic colitis, and not significantly more frequent than in controls in some studies [32, 33]. When found, titres of these antibodies in microscopic colitis are lower than in coeliac patients [32]. These data support the conclusion that sprue is uncommon in patients with microscopic colitis. It may not be necessary to routinely evaluate patients with microscopic colitis for sprue, but this association should be considered in treatment-refractory patients, those with significant weight loss or any suggestion of steatorrhoea, or other clues such as unexplained iron-deficiency anaemia.

Endoscopic evaluation of the colon is typically normal or has mild nonspecific changes such as erythema or oedema. Colonic ulceration is uncommon, and when seen is likely related to use of non-steroidal anti-inflammatory drugs (NSAIDs) [34].

The key histologic finding in microscopic colitis is intra-epithelial lymphocytosis [35, 36]. In addition, there is a mixed inflammatory infiltrate in the lamina propria [36]. These inflammatory changes are often accompanied by surface epithelial damage [36]. In collagenous colitis, the subepithelial collagen band is abnormally thickened, compared with 5–7 μm in normals [36].

Pathophysiology

Data on the mechanisms involved in microscopic colitis come mostly from small studies, and no consistent mechanism has been established [37]. Postulated mechanisms have included bile acid malabsorption, altered fluid and electrolyte absorption or secretion, unidentified infection, immunologic reaction to a luminal antigen (food, microorganism, or other), autoimmunity, hormonal influence (given female predominance) and alteration in collagen synthesis or degradation (in collagenous colitis). Thus, it is likely that the clinicopathologic term 'microscopic colitis' encompasses several different aetiologies or pathophysiologic mechanisms with similar clinical and histologic phenotypes.

One postulated pathophysiologic mechanism that has clinical relevance is the entity of drug-induced microscopic colitis [38, 39]. Several drugs have been implicated as possible causes of microscopic colitis, including NSAIDs, histamine-2 receptor blockers, proton pump inhibitors, selective serotonin re-uptake inhibitors, carbamazepine, simvastatin, ticlopidine, and others (table 1) [37–39]. In some cases, symptoms and histologic changes resolved with drug withdrawal and returned with re-exposure. However, for most drugs, re-challenge is not reported, and the number of cases is small, such that a chance association cannot be excluded. Furthermore, some drugs implicated in causing microscopic colitis are also associated with watery diarrhoea, and therefore they may not actually cause colitis, but rather worsen the diarrhoea and thus bring the diagnosis to attention [39].

Table 1. Drugs and drug classes implicated as causing microscopic colitis, listed according to level of evidence in support of causality. Adopted from [38]

High level evidence
Acarbose
Aspirin
Nonsteroidal anti-inflammatory drugs
Proton pump inhibitors
Selective serotonin reuptake inhibitors
Ticlopidine
Intermediate level evidence
Carbamazepine
Flutamide
Lisinopril
Simvastatin

Treatment

In the past, treatment of microscopic colitis was largely based on observational uncontrolled reports and anecdotal evidence on a variety of drugs such as antidiarroeals, bulking agents, bismuth subsalicylate, corticosteroids, budesonide, 5-ASA compounds, bile acid binding agents, various immunosuppressives, octreotide, antibiotics, and even surgery. Interpretation of theses data have been hampered by the variable course of disease including a considerable rate of spontaneous remission, stressing the need for randomized controlled studies. Moreover, there are no standardized or generally accepted definitions for clinical (and histological) response or remission, and the outcome measures for defining these criteria vary between trials. Fortunately, a number of randomized trials, mostly in collagenous colitis, have been performed and published in recent years, providing important data for a more evidence-based and rational approach to treatment of patients with microscopic colitis. The data for the various therapies for microscopic colitis are reviewed below, with a particular focus on randomized controlled clinical trials, where they exist. Additional comments on uncontrolled treatment reports can be found elsewhere [37]. A summary of treatment response in 3 large open-label series is found in table 2.

Budesonide

The strongest evidence from clinical trials for the treatment of microscopic colitis is currently available for budesonide, a locally active corticosteroid with an extensive first-pass metabolism in the liver and a low systemic exposure. Three short-

Table 2. Response to therapy in 3 open-label series for microscopic colitis

	Pardi et al. [25] (n = 170)	Olesen et al. [14] (n = 199)	Bohr et al. [12] (n = 163)
Type of colitis	Lymphocytic	Lymphocytic	Collagenous
% of subjects responding to			
Antidiarrheals	73	70	71
Bismuth	73		
Sulfasalazine	42	21	34
Mesalamine	45	50	50
All 5-ASA	42	37	35
Cholestyramine	65	57	59
Steroids	87	88	82
AZA/6-MP	20		

Data include complete and partial responders. Response rates represent clinical experience and are not the result of controlled trials. ASA = Aminosalicylates; AZA = azathioprine; 6-MP = 6-mercaptopurine.

term randomized, double-blind, placebo-controlled trials in collagenous colitis that included a total of 94 patients have consistently demonstrated that budesonide 9 mg daily for 6–8 weeks is effective and superior to placebo [40–42]. A Cochrane meta-analysis revealed a pooled response rate of 81% (compared to 17% with placebo), a pooled odds ratio for clinical response of 12.32, and a number needed to treat of 2 patients [43]. Budesonide was generally well tolerated and also improved histology and the patient's quality of life [24, 43]. However, 60–80% of patients may suffer from symptomatic relapse after cessation of short-term treatment [40, 44]. Two randomized, double-blind, placebo-controlled trials have now shown that budesonide 6 mg per day for 6 months is well tolerated and effective in maintaining clinical response in patients with collagenous colitis, with a pooled response rate of 83% (compared to 28% with placebo) [45, 46]. The pooled odds ratio was 8.4, with a number needed to treat of 2 patients for maintaining clinical response [43]. Budesonide was also superior to placebo in maintaining histological response. However, the risk of symptomatic relapse appears not to be decreased after cessation of budesonide maintenance therapy [46].

The beneficial effect of budesonide 9 mg per day for 6 weeks has also been confirmed in lymphocytic colitis by 2 randomized, double-blind, placebo-controlled trials including 41 and 15 patients [47, 48]. In one study, budesonide was superior to placebo, with clinical remission rates of 86 versus 40% [47]. A Cochrane meta-analysis revealed an odds ratio of 9 and a number needed to treat of 3 patients for clinical response with budesonide [49]. The second study showed similar results, with clinical response rates of 91% in budesonide patients and 25% in placebo patients [48]. Both studies also demonstrated improvement in histology.

Prednisolone

Prednisolone has been investigated in only one randomized, placebo-controlled trial, which included 11 patients with collagenous colitis and 1 with lymphocytic colitis [50]. Clinical response was noted in 5 of 8 patients after 2 weeks of prednisolone compared to none of 3 patients with placebo. Due to the small number of patients, the study lacked power to detect a significant difference. Adverse events were common in prednisolone patients. Furthermore, the effect of prednisolone on histology has not been studied. Interpretation of the results is limited by the small number of patients and the short treatment period. Thus, larger trials would be necessary to assess the efficacy of systemic prednisolone in patients with collagenous colitis.

Bismuth Subsalicylate

One randomized placebo-controlled trial, published as abstract only, studied bismuth subsalicylate 3 × 262 mg given orally t.i.d. for 8 weeks in 9 patients with collagenous colitis and 5 patients with lymphocytic colitis [51]. Clinical and histological response was noted in all 7 bismuth subsalicylate patients compare to none of 7 placebo recipients. No adverse events have been reported. However, due to concerns regarding toxicity, this drug is not available in a number of countries.

Mesalazine with or without Cholestyramine

In one randomized study, 64 patients (23 with collagenous colitis, 41 with lymphocytic colitis) received mesalazine 2.4 g per day alone or in combination with cholestyramine 4 g per day for 6 months [52]. Remission occurred in 91% of patients with collagenous colitis and 85% of patients with lymphocytic colitis after 6 months, and combined treatment appeared to be slightly better in collagenous colitis. Weaknesses of this study include the lack of blinding and the lack of a placebo control group.

Probiotics

The probiotics *Lactobacillus acidophilus* LA-5 and *Bifidobacterium animalis subspp. lactis* BB12 (AB-Cap-10) given for 12 weeks did not show any benefit over placebo in a randomized, double-blind, placebo-controlled study in 29 patients with collagenous colitis [53]. Clinical response was noted in 6 of 21 probiotic patients and in 1 of 8 placebo recipients. There was no effect on histology or on quality of life.

Boswellia serrata Extract

In a randomized, double-blind, placebo-controlled trial, *Boswellia serrata* extract 3 × 400 mg per day for 6 weeks failed to show a significant benefit over placebo with clinical responses in 7 of 16 patients (44%) compared to 4 of 15 placebo patients (27%). Neither histology nor quality of life was improved by active treatment [54].

Antidiarrhoeals

Antidiarrhoeals such as loperamide are frequently used as first-line treatment in patients with collagenous or lymphocytic colitis, although their benefit has never been strictly tested in a randomized placebo-controlled trial. However, from clinical experience, a symptomatic benefit of antidiarrhoeals may be seen in a considerable proportion of patients. Therefore antidiarrhoeals may be used alone or in conjunction with other therapies depending on symptom severity [12, 14, 25].

Immunosuppressive Therapy

Although the evidence is limited, immunosuppressive therapies may be considered in patients with severe symptoms who fail to respond to steroids or who frequently relapse when tapering or discontinuing steroids. In such cases, azathioprine or 6-mercaptopurine and methotrexate could be used in dosages similar to those used for other inflammatory bowel diseases [55, 56].

Surgical Therapy

The indications for surgical intervention in microscopic colitis should be very limited given the substantial improvement of medical therapies. Both diverting ileostomy and subtotal colectomy have been performed successfully in individual cases [57, 58]. Therefore, surgical therapy could be considered in patients with severe microscopic colitis who fail all appropriate medical therapies.

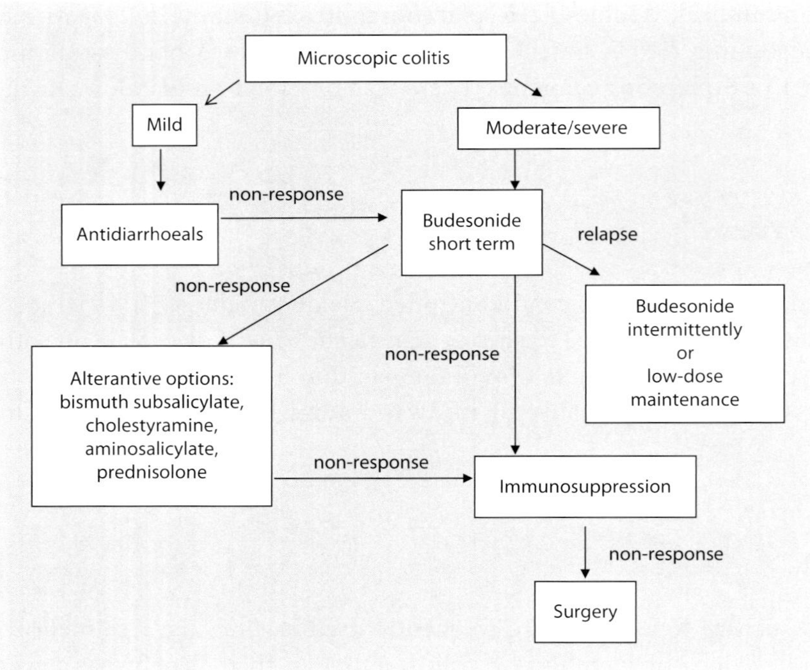

Fig. 1. A treatment algorithm for microscopic colitis.

Conclusion

Microscopic colitis is an increasingly common cause of chronic diarrhoea. Colonic biopsies are required to make the diagnosis, and should be performed in all patients undergoing sigmoidoscopy or colonoscopy for unexplained diarrhoea. The 2 subtypes of microscopic colitis, collagenous and lymphocytic colitis, are similar histologically and clinically, and seem to respond similarly to various medical therapies.

Decisions on medical therapy of microscopic colitis should be based on the severity and course of the patient's symptoms, the impact on patient's quality of life, and the response to previous treatments. In addition, the potential for concomitant coeliac disease or drug-induced microscopic colitis should be taken into account before embarking on medical therapy. Finally, the availability of evidenced-based or approved treatment modalities should also be considered.

Patients with only mild intermittent symptoms may be successfully managed with antidiarrhoeals. Some clinicians also use bismuth subsalicylate in these patients, although the evidence supporting this practice is limited. Those with moderate to severe symptoms, which are often disabling and associated with considerable impairment of quality of life, should receive short-term budesonide as first-line treatment. In case of recurrent disease, budesonide may be used again intermittently or as low-dose maintenance therapy [59]. With long-term budesonide treatment, patients should be monitored for steroid-related side effects. In patients who do not respond to budesonide, other drugs such as bismuth subsalicylate (when available), cholestyramine, mesalazine or prednisolone/prednisone may be considered, depending on symptom severity. In refractory cases, immunosuppressive agents such as azathioprine, 6-mercaptopurine or methotrexate may be useful. Surgery may be considered in patients not responding to any medical therapy. A treatment algorithm for patients with microscopic colitis is outlined in figure 1.

References

1 Read NW, Krejs GJ, Read MG, et al: Chronic diarrhea of unknown origin. Gastroenterology 1980;76: 264–271.
2 Lindstrom CG: 'Collagenous colitis' with watery diarrhoea: a new entity? Pathologica Europa 1976; 11:87–89.
3 Jessurun J, Yardley JH, Lee EL, et al: Microscopic and collagenous colitis: different names for the same condition? Gastroenterology 1986;91:1583–1584.
4 Levison DA, Lazenby AJ, Yardley JH: Microscopic colitis cases revisited. Gastroenterology 1993;105: 1594–1596.
5 Fine KD, Seidel RH, Do K: The prevalence, anatomic distribution, and diagnosis of colonic causes of chronic diarrhea. Gastrointest Endosc 2000;51: 318–326.
6 Shah RJ, Bleau B, Giannella RA: Usefulness of colonoscopy with biopsy in the evaluation of patients with chronic diarrhea. Am J Gastroenterol 2001;96:1091–1095.
7 Fernandez-Banares F, Salas A, Forne M, et al: Incidence of collagenous and lymphocytic colitis: a 5-year population-based study. Am J Gastroenterol 1999;94:418–423.
8 Pardi DS, Loftus EV, Smyrk TC, et al: The epidemiology of microscopic colitis: a population-based study in Olmsted County, MN. Gut 2007;56:504–508.
9 Bohr J, Tysk C, Eriksson S, et al: Collagenous colitis in Orebro, Sweden, an epidemiological study 1984–1993. Gut 1995;37:394–397.
10 Agnarsdottir M, Gunnlaugsson O, Orvar KB, et al: Collagenous and lymphocytic colitis in Iceland. Dig Dis Sci 2002;47:1122–1128.
11 Williams JJ, Kaplan GG, Makhija S, et al: Microscopic colitis-defining incidence rates and risk factors: a population-based study. Clin Gastroenterol Hepatol 2008;6:35–40.
12 Bohr J, Tysk C, Eriksson S, et al: Collagenous colitis: a retrospective study of clinical presentation and treatment in 163 patients. Gut 1996;39:846–851.
13 Fernandez-Banares F, Salas A, Esteve M, et al: Collagenous colitis and lymphocytic colitis: evaluation of clinical and histological features, response to treatment and long-term follow up. Am J Gastroenterol 2003;98:340–347.
14 Olesen M, Erickson S, Bohr J, et al: Lymphocytic colitis: a retrospective study of 199 Swedish patients. Gut 2004;53:536–541.
15 Gremse DA, Boudreaux CW, Manci EA: Collagenous colitis in children. Gastroenterology 1993;104:906–909.
16 Mahajan L, Wyllie R, Goldblum J: Lymphocytic colitis in a pediatric patient: a possible adverse reaction to carbamazepine. Am J Gastroenterol 1997;92: 2126–2127.
17 Jarnerot G, Hertervig E, Granno E, et al: Familial occurrence of microscopic colitis: a report of five families. Scand J Gastroenterol 2001;36:959–962.
18 Abdo AA, Zetler PJ, Halparin LS: Familial microscopic colitis. Can J Gastroenterol 2001;15:341–343.
19 Chan JL, Tersmette AC, Offerhaus GJA, et al: Cancer risk in collagenous colitis. Inflamm Bowel Dis 1999;5:40–43
20 Pardi DS, Kammer PP, Loftus EV, et al: Microscopic colitis is not associated with an increased risk of colorectal cancer (abstract). Am J Gastroenterol 2004; 99:S115.
21 Ung KA, Kilander A, Willen R, et al: Role of bile acids in lymphocytic colitis. Hepatogastroenterology 2002;49:432–437.
22 Pardi DS, Loftus EV, Kammer PP, et al: Collagenous colitis, but not lymphocytic colitis, is associated with cigarette smoking (abstract). Am J Gastroenterol 2004;99:S116.

23 Mullhaupt B, Guller U, Anabitarte M, et al: Lymphocytic colitis: clinical presentation in long term course. Gut 1998;43:629–633.

24 Madisch A, Heymer P, Voss C, et al: Oral budesonide therapy improves quality of life in patients with collagenous colitis. Int J Colorectal Dis 2005;20:312–316.

25 Pardi DS, Ramnath VR, Loftus EV, et al: Lymphocytic colitis: clinical features, treatment, and outcomes. Am J Gastroenterol 2002;97:2829–2833.

26 Limsui D, Pardi DS, Camilleri M, et al: Symptomatic overlap between irritable bowel syndrome and microscopic colitis. Inflamm Bowel Dis 2007;13:175–181.

27 Madisch A, Bethke B, Stolte M, Miehlke S: Is there an association of microscopic colitis and irritable bowel syndrome: a subgroup analysis of placebo-controlled trials. World J Gastroenterol 2005;11:6409.

28 Bohr J, Tysk C, Yang P, et al: Autoantibodies and immunoglobulins in collagenous colitis. Gut 1996;39:73–76.

29 Wolber R, Owen D, Freeman H: Colonic lymphocytosis in patients with celiac sprue. Hum Pathol 1990;21:1092–1096.

30 Breen EG, Coughlan G, Connolly CE, et al: Coeliac proctitis. Scand J Gastroenterol 1987;22:471–477.

31 Fine KD, Meyer RL, Lee EL: The prevalence and cause of chronic diarrhea in patients with celiac sprue treated with a gluten-free diet. Gastroenterology 1997;112:1830–1838.

32 Fine KD, Do K, Schulte K, et al: High prevalence of celiac sprue-like HLA-DQ genes and enteropathy in patients with the microscopic colitis syndrome. Am J Gastroenterol 2000;95:1974–1982.

33 Ramzan NN, Shapiro MS, Pasha TM, et al: Is celiac disease associated with microscopic and collagenous colitis (abstract)? Gastroenterology 2001;120:A684.

34 Kakar S, Pardi DS, Burgart LJ: Colonic ulcers accompanying collagenous colitis: implication of NSAIDs. Am J Gastroenterol 2003,98:1834–1837.

35 Baert F, Wouters K, D'Haens G, et al: Lymphocytic colitis: a distinct clinical entity? A clinicopathological confrontation of lymphocytic and collagenous colitis. Gut 1999;45:375–381.

36 Lazenby AJ, Yardley JH, Giardiello FM, et al: Lymphocytic (microscopic) colitis: a comparative histopathologic study with particular reference to collagenous colitis. Hum Pathol 1989;20:18–28.

37 Pardi DS: Microscopic colitis: an update. Inflamm Bowel Dis 2004;10:860–870.

38 Beaugerie L, Pardi DS: Drug-induced microscopic colitis: proposal for a scoring system and review of the literature. Aliment Pharmacol Ther 2005;22:277–284.

39 Fernández-Bañares F, Esteve M, Espinós JC, et al: Drug consumption and the risk of microscopic colitis. Am J Gastroenterol 2007;102:324–330.

40 Baert F, Schmit A, D'Haens G, et al: Budesonide in collagenous colitis: a double-blind placebo-controlled trial with histologic follow-up. Gastroenterology 2002;122:20–25.

41 Miehlke S, Heymer P, Bethke B, et al: Budesonide treatment for collagenous colitis: a randomized, double-blind, placebo-controlled, multicenter trial. Gastroenterology 2002;123:978–984.

42 Bonderup OK, Hansen JB, Birket-Smith L, et al: Budesonide treatment of collagenous colitis: a randomised, double-blind, placebo-controlled trial with morphometric analysis. Gut 2003:52:248–251.

43 Chande N, McDonald JW, McDonald JK: Interventions for treating collagenous colitis. Cochrane Database Syst Rev 2008;16:CD003575.

44 Miehlke S, Madisch A, Voss C, et al: Long-term follow-up and predictive factors for clinical relapse in patients with collagenous colitis after induction of remission with budesonide capsules. Aliment Pharmacol Ther 2005;22:1115–1119.

45 Bonderup OK, Hansen JB, Teglbjaerg PS, et al: Long-term budesonide treatment of collagenous colitis: a randomised, double-blind, placebo-controlled trial. Gut 2009;58:68–72.

46 Miehlke S, Madisch A, Bethke B, et al: Budesonide for maintenance treatment of collagenous colitis: a randomised, placebo-controlled, double-blind trial. Gastroenterology 2008;135:1510–1516.

47 Miehlke S, Madisch A, Karimi D, et al: Budesonide is effective in treating lymphocytic colitis: a randomized double-blind placebo-controlled study. Gastroenterology 2009;136:2092–2100.

48 Pardi DS, Loftus EV, Tremaine WJ, Sandborn WJ: A randomized, double-blind, placebo-controlled trial of budesonide for the treatment of active lymphocytic colitis. Gastroenterology 2009;136(suppl 1):T1193.

49 Chande N, McDonald JW, MacDonald JK: Interventions for treating lymphocytic colitis. Cochrane Database Syst Rev 2008;16:CD006096.

50 Munck LK, Kjeldsen J, Philipsen E, et al: Incomplete remission with short-term prednisolone treatment in collagenous colitis: a randomized study. Scand J Gastroenterol 2003;38:606–610.

51 Fine K, Ogunji F, Lee E, et al: Randomized, double-blind, placebo-controlled trial of bismuth subsalicylate for microscopic colitis. Gastroenterology 1999;116:A880.

52 Calabrese C, Fabbri A, Areni A, et al: Mesalazine with or without cholestyramine in the treatment of microscopic colitis: randomized controlled trial. J Gastroenterol Hepatol 2007;22:809–814.

53 Wildt S, Munck LK, Vinter-Jensen L, et al: Probiotic treatment of collagenous colitis: a randomized, double-blind, placebo-controlled trial with *Lactobacillus acidophilus* and *Bifidobacterium animalis subsp. lactis*. Inflamm Bowel Dis 2006;12:395–401.
54 Madisch A, Miehlke S, Eichele O, et al: *Boswellia serrata* extract for the treatment of collagenous colitis: a double-blind, randomized, placebo-controlled, multicenter trial. Int J Colorectal Dis 2007;22:1445–1451.
55 Pardi DS, Loftus EV, Tremaine WJ, Sandborn WJ: Treatment of refractory microscopic colitis with azathioprine and 6-mercaptopurine. Gastroenterology 2001;120:1483–1484.
56 Riddell J, Hillman L, Chiragakis L, Clarke A: Collagenous colitis: oral low-dose methotrexate for patients with difficult symptoms: long-term outcomes. J Gastroenterol Hepatol 2007;22:1589–1593.
57 Jarnerot G, Tysk C, Bohr J, Eriksson S: Collagenous colitis and fecal stream diversion. Gastroenterology 1995;109:449–455.
58 Munch A, Söderholm JD, Wallon C, et al: Dynamics of mucosal permeability and inflammation in collagenous colitis before, during, and after loop ileostomy. Gut 2005;54:1126–1128.
59 Pardi DS: After budesonide, what next for collagenous colitis? Gut 2009;58:3–4.

Darrell S. Pardi
Inflammatory Bowel Disease Clinic, Division of Gastroenterology and Hepatology
Mayo Clinic College of Medicine
Rochester, Minnesota (USA)
Tel. +1 507 538 1231, Fax +1 507 284 0538, E-Mail pardi.darrell@mayo.edu

Mayerle J, Tilg H (eds): Clinical Update on Inflammatory Disorders of the Gastrointestinal Tract.
Front Gastrointest Res. Basel, Karger, 2010, vol 26, pp 146–156

Inflammatory Proteins as Prognostic Markers in Acute Pancreatitis

Jean Louis Frossard[a] · Madhav Bhatia[b]

[a]Division of Gastroenterology, Geneva University Hospitals, Geneva, Switzerland; [b]Cardiovascular Biology Research Programme, Life Sciences Institute, Department of Pharmacology, NUS Graduate School for Integrative Sciences and Engineering, Centre for Life Sciences, National University of Singapore, Singapore

Abstract

Acute pancreatitis (AP) is an inflammatory disease of the pancreas, which involves a gland that was normal prior to the onset of the disease and which may return to normality after disease resolution. Acute abdominal pain is the most common presenting symptom of the disease and increased serum amylase and lipase concentrations confirm the diagnosis. Pancreatic injury is mild in 80% of patients, who will recover without complication. The remaining 20% of patients have severe disease. Gallstone migration into the common bile duct and alcohol abuse are the most frequent aetiologies of the disease. Because it is important to predict the severity of the illness as early as possible in order to optimize the therapy and to prevent organ dysfunction and local complications, several clinical scores of severity have been proposed. These scores assess the multiple organ dysfunction induced by the disease and consequently, the greater the number of organs injured, the greater the score. However, the complexity of the scales has prompted some authors to evaluate inflammatory proteins as prognostic markers in acute pancreatitis, including C-reactive protein, and cytokines and chemokines. This paper is aimed at reviewing data that cover the field combining prediction of pancreatitis severity and inflammatory proteins. A better understanding of the pathogenesis as well as a better assessment of the disease severity should further improve the management and outcome of the disease.

Acute pancreatitis (AP) is a common disease of variable severity in which some patients experience mild symptoms while others present with severe, sometimes lethal attacks [1, 2]. In about 80% of patients, AP is mild and resolves without serious morbidity, but in the remaining 20% it is complicated by significant morbidity and high mortality. Between 10 and 15% of patients with severe disease will die from it [2–4]. The incidence of AP has increased in the past two decades [4–8]. This increased incidence probably reflects greater accuracy in the diagnosis of the disease.

The epidemiology of AP is determined by its aetiology. In women, biliary stone migration is more frequent and alcohol abuse is less common than in men [9, 10].

AP in the first two decades of life suggests a genetic cause. Although there are a great number of aetiologies of AP, for 75–85% of patients the cause is easily identified. In Western countries, gallstone migration in the common bile duct is the most frequent cause of AP (38%) while alcohol abuse (36%) is the second most frequent cause [11]. In addition to less frequent aetiologies – such as drug-induced, autoimmune or metabolic pancreatitis – about 15–25% of pancreatitis episodes are of an unknown origin. Recently, mutations of the cationic trypsinogen gene ($PRSS_1$), serine protease inhibitor Kazal type 1 (SPINK1) and cystic fibrosis transmembrane conductance regulator (CFTR) genes have been detected in patients with idiopathic recurrent AP.

Insights into Pathophysiology

Recent advances have improved our understanding of the pathophysiology of AP. Indeed, the emerging concept for the past 10 years is that AP includes the premature activation and release of pancreatic enzymes in the interstitium, the autodigestion of the pancreas by these enzymes and the multiple organ dysfunction following the release of several mediators in the systemic circulation. Indeed, AP occurs when intracellular protective mechanisms designed to prevent trypsinogen activation or reduce trypsin activity do not function fully. These protective mechanisms include the synthesis of trypsin as the inactive enzyme trypsinogen, enzyme compartmentalization and packaging, synthesis of specific trypsin inhibitors such as serine protease inhibitor Kazal type 1 ($SPINK_1$) and low intracellular calcium concentrations. During AP, following the activation of trypsinogen into active trypsin within acinar cells, numerous enzymes such as elastase and phospholipase A_2 as well as complement and kinin systems are activated [12]. Additionally, inflammation is initiated with local production of mediators such as IL-1, IL-6 and IL-8 from neutrophils, macrophages and lymphocytes. TNF-α is also released by local macrophages within pancreatic tissue and its production correlates with the severity of the experimental disease [13]. Interestingly, anti-inflammatory cytokines, such as IL-10 decrease the severity of experimental pancreatitis [14].

Over the last few years, significant evidence has been accumulated that this multisystem pathological condition is caused by the synthesis and release of pro-inflammatory cytokines and chemokines. These pro-inflammatory mediators are thought to be produced within the gland. They can amplify localized pancreatic injury and then translate local inflammation into systemic disease with the involvement of distant organ [15–17].

In addition to cytokine and chemokine synthesis and release, activation of endothelial cells permits the transendothelial migration of leukocytes that release other harmful enzymes [18]. Decreased O_2 delivery to the organ and generation of O_2-derived free radicals also contribute to the injury [19]. Thus, regardless of the initial trigger of the disease, the severity of pancreatic damage is related to the injury of acinar cells and to the activation of inflammatory and endothelial cells. Local complications (acinar cell necrosis, pseudocyst formation and abscess) and injury in remote organs (lungs) follow the release of numerous mediators by these cells [20].

Recent studies have also shown a key role played by substance P [21, 22], a neuropeptide, and hydrogen sulphide (H_2S) [23, 24], a gas, in the pathogenesis of AP and systemic organ damage.

Severe attacks of AP are frequently associated with acute lung injury. Complications of AP include arteriolar hypoxaemia, pulmnonary infiltrates and pleural effusions, as well as acute respiratory distress syndrome.

Prediction of Severity

Clinical findings are unreliable to determine the prognosis of AP and so prognostic indicators of all types are being increasingly used. Since treatment of patients with severe disease is currently mostly supportive and directed at either prevention or control of the local and systemic complications, the early prediction of the severity of an attack of AP is the main goal of clinicians in charge of such patients. Because it is important to predict the severity of the illness as early as possible in order to optimize the therapy and to prevent organ dysfunction and local complications, several scores of severity have been proposed. Criteria of severity, such as Ranson [25, 26], Glasgow [27] and the Acute Physiology and Chronic Health Evaluation (APACHE) [28] scores have been used for a long time. These scores assess the multiple organ dysfunction induced by the disease and, consequently, the greater the number of organs injured, the greater the score.

The Ranson scale has been used since the 1980s in virtually all studies dealing with AP. Unfortunately, the list of 11 numerical parameters proposed by Ranson requires little discussion [29], because only 5 criteria are obtained at admission while the remaining 6 are assessed in the subsequent 48 h, making it impossible to make an early prediction. In 1981, Knaus et al. [69] proposed the APACHE disease grading system, which assigned 5 possible grades to 34 physiological and laboratory determinations, the sum of which represented the acute physiology score. A simplified version containing only 12 determinations was proposed in 1985 (Apache II), but because this multifactorial scoring system is complex and requires additional data frequently unavailable outside intensive care units, it remains inconvenient to perform [30].

Inflammatory Proteins as Prognostic Markers

The complexity of the Ranson and APACHE II scales has prompted some authors to evaluate inflammatory proteins as prognostic markers in AP. New biological criteria have recently emerged and their ability to provide additional information on the severity of the disease is currently being evaluated [31, 32]. Upon admission, numerous biological mediators can be detected in blood samples. If the concentration of these biological factors is correlated to the severity of the disease and if they are detected before the occurrence of multiple organ dysfunction, it is conceivable that these new factors might be important for the rapid scoring of the disease severity in the acute phase.

C-Reactive Protein

C-Reactive protein (CRP) responds to any inflammatory challenge within 4–6 hours and is primarily synthesized by hepatocytes. CRP was first used as a prognostic factor in the assessment of pancreatitis severity in the mid 1980s and has since gained undisputed recognition, with most authors suggesting that a 24- to 48-hour latency is necessary. In the study by Buchler [32], it has given CRP prominent importance in clinical decision-making, since values greater than 100 mg/l were found to carry a 95% detection rate for pancreatic necrosis. In the studies by Wilson et al. [33] and Leese et al. [34], similar views were shared about the usefulness of CRP in predicting severe outcome. Values greater than 120 mg/l can detect between 67 and 100% of pancreatic necrosis [35]. Neutrophil elastase on admission and CRP the following day were shown to be the most predictive variables in the outcome of AP in 80 non-consecutive patients [36]. Duarte-Rojo et al. [37] found that in patients admitted during the first 48 h, IL-6 and IL-10 had improved accuracy over the Ranson score, whereas after this time frame, only CRP outperformed Ranson. Therefore, CRP can be considered as a valuable tool compared to the complex multiple factor scoring systems used in clinical practice. However, because a 24- to 48-hour latency is necessary before detecting a CRP increase in plasma, other inflammatory mediators, such as cytokines or chemokines, have been suggested [38, 39].

Cytokines

Interleukin-1

IL-1 is a pro-inflammatory mediator synthesized and released during acute inflammation and septic conditions. McKay et al. [39] studied the release of cytokines from monocytes isolated from the peripheral blood of patients with moderate and severe diseases between days 1 and 7. Although TNF-α, IL-6 and IL-8 release were significantly higher in patients with systemic complications than in those with uncomplicated disease, the IL-1 levels were similar in the 2 groups. The results obtained from 37 patients (25 severe and 12 mild cases) confirm that CRP and IL-6 determinations within 48 hours after admission predict the severity of the attack more accurately than IL-1. Although associated with the patient's severity at inclusion and outcome, cytokine plasma concentrations, including IL-1 levels, were unable to predict death accurately in individual patients [41].

Interleukin-6

Since 1990, several new prognostic factors for assessing the severity of AP have been published. Among them, IL-6 is of great interest in AP. In AP, IL-6 represents a very early marker of severe outcome and some authors have already introduced an IL-6 blood level rapid measurement system, which can evaluate the IL-6 blood level within 30 min [42]. In the study by Gross et al. [35], maximal IL-6 concentrations correlated with maximal increase in the serum concentrations of CRP. Moreover, patients with lethal outcome had markedly elevated initial IL-6 concentrations [43]. These findings were then confirmed by Watanabe's group showing that IL-6 is able to distinguish between severe and mild attacks of AP with a sensitivity of 100% and a specificity of 71% [44]. Brivet et al. [40] showed that hospital mortality was linked to 6 factors on univariate analysis (age, cirrhosis, delay between hospitalization and intensive care unit admission, severity of illness, and IL-10 and IL-6 plasma levels), but using stepwise logistic regression only severity scoring indexes were predictive of death.

Recently, the CC genotype of the IL-6 174 polymorphism has been shown to be associated with a biliary aetiology of AP [44]. Watanabe et al. [45] found that excessive activation of the cytokine network by the TNF-308*A, IL1RN*2, and RN*3 alleles was involved in the mechanism of onset and poor outcome in patients with extremely high IL-6 blood level.

Chemokines and Neutrophil Elastase

Since it has been suggested that neutrophils play a key role in the pathogenesis of AP [15], IL-8, a potent neutrophil chemoattractant agent, has been shown to be higher in patients with severe disease than those with mild cases. Moreover, IL-8 levels were positively correlated with the neutrophil elastase concentrations reflecting the amount of neutrophils activated. When measured during the first 24 h after disease onset, IL-6 and IL-8 levels were shown to be better markers than CRP for evaluating the severity of the disease [46]. In another study, the chemokines epithelial-derived neutrophil-activating peptide (ENA)-78 and growth-related oncogene (Gro)-α, in addition to IL-8, were shown to be predictors of the severity of AP [47]. Polymorphisms of the IL-8 gene were recently studied in patients with severe pancreatitis. The IL-8 A/T heterozygote mutant variants were detected with a significantly higher frequency among patients with severe pancreatitis than among healthy blood donors, while the frequency of the normal allelic genotype TT was higher among patients with mild pancreatitis than in the group with severe pancreatitis [48]. Neutrophil elastase has also been shown to be an early marker of outcome in AP. Indeed, neutrophil elastase reached significantly higher values in severe cases than in mild attacks as a reflection of considerable leucocyte activation [49]. With a positive predictive value of >97% 24 h after admission, neutrophil elastase seems to be a very reliable marker in the diagnosis of severe disease [50].

Interleukin-10

IL-10 recently emerged as an anti-inflammatory cytokine which inhibits the secretion of pro-inflammatory cytokines by monocytes/macrophages. Experiments in animals have shown that knock-out mice for IL-10 had worse pancreatitis than wild-type and that treatment with IL-10 decreases the severity of experimental pancreatitis [51, 52]. Not surprisingly, serum IL-10 levels were found to be lower in patients with severe disease than in mild cases, but only within the first 24 h [53]. However, in another study, IL-10 levels were higher when AP was severe than when the injury was mild [54]. The same trend was found by Stimac et al. [55] recently who showed that IL-10 did not properly assess complicated versus non-complicated AP.

Other Cytokines

While Th1 type cells produce high levels of inflammatory cytokines including IL-2, TNF-α, and IFN-γ, Th2 type cells produce anti-inflammatory cytokines including IL-4, IL-5, IL-6, IL-10, and IL-13. Interleukin-2 and Interleukin-11 do not constitute an early potent marker of the severity of AP [56]. In patients with AP, IL-12 levels were significantly higher from days 1 to 6 than in healthy subjects [57]. IL-15 is a novel cytokine that shares many biologic properties with IL-2. In one study, serum IL-15 levels increased significantly in severe AP and correlated with Ranson and APACHE II scoring systems. Interestingly, the usefulness of IL-15 for the prediction of organ dysfunction was superior to CRP, IL-6, and IL-8, but was similar to Ranson and APACHE II [58].

TNF-α is an early mediator of sepsis and organ failure and a potent inducer of IL-6 secretion, suggesting that TNF-α could be of great interest in pancreatitis. However, in contrast with CRP and IL-6, TNF-α was not identified as a valuable tool in assessing the severity of AP. The action of TNF-α on an organ depends on the presence of specific receptors on the surface of the recipient cells (TNFR). Elevated plasma concentrations of TNFR appear to better reflect TNF-α induced inflammation. In the study by Kaufman et al. [60], TNFR shows a sensitivity and a specificity of 90 and 100%, respectively, in predicting the severity of the disease.

Intercellular adhesion molecule-1 (ICAM-1) is an inducible membrane glycoprotein expressed by leucocytes, fibroblasts and endothelial cells. It plays a key role in adhesion of neutrophils to the endothelium of microvessels and their transendothelial migration. Soluble isoforms of ICAM-1 result from shedding of transmembrane ICAM-1 and are elevated during inflammation. The sensitivity and specificity for the detection of oedematous or necrotizing pancreatitis of sICAM-1 were 75 and 85% respectively within 24–28 h of onset of the disease [61].

Other Markers

TREM-1

Triggering receptor expressed on myeloid cells (TREM)-1, a new receptor of the immunoglobulin superfamily identified recently, is expressed on neutrophils and monocytes/macrophages [62]. The usefulness of serum sTREM-1 in detecting early organ dysfunction was superior to that of CRP, IL-6, IL-8, Ranson score, and APACHE II score. Serum sTREM-1 levels decreased with resolution of early organ dysfunction [63].

Phospholipase

At the site of inflammation, activated neutrophils release numerous active substances such as phospholipase A which can be used as a phagocytic marker in inflammation and necrosis. In the study by Viedma et al. [64], IL-6 levels correlated both with clinical severity and with CRP and phospholipase A.

Procalcitonin

Recently, procalcitonin, a 116 amino acid propeptide of calcitonin, was found to appear in high concentrations in patients with severe bacterial or fungal infections and sepsis. Procalcitonin is usually not detectable in healthy controls. In a prospective study of 50 patients with AP, median concentrations of procalcitonin were significantly higher in patients with infected necrosis than in those with sterile necrosis whereas there was no difference in CRP levels. It therefore seems that procalcitonin might be a valuable tool for the non-invasive and accurate prediction of infected necrosis.

Neopterin

Neopterin, a seric marker of macrophage activation, has also been studied in patients with AP. Serum neopterin concentrations were higher in severe than in mild pancreatitis at day 1 [65]. However, neopterin serum values did not correlate with IL-6 and TNF-α concentrations at any day. In the study by Mora et al. [66]., neopterin was not superior to neutrophil elastase in predicting the severity of the disease. In a prospective study of 25 patients with AP [67], serum neopterin concentrations correlated with the severity of the disease determined by the APACHE II score. In this study, the discrimination between mild and severe pancreatitis was higher for neopterin than for CRP. Although neopterin seems to be a good marker, its measurement is not available in most institutions.

Hydrogen Sulphide

In support of results obtained with preclinical studies using experimental models [23, 24], an early clinical study suggests that H_2S is involved in the inflammatory response of AP in humans. Serum H_2S levels within 24 hours correlated with subsequent severity of AP, suggesting a potential role of H_2S as an early predictor of disease severity [68].

Table 1. Inflammatory mediators that have been identified as prognostic markers in acute pancreatitis

C-reactive protein
Interleukins 1, 6, 8, 10, 12, 15
Growth-related oncogene-α
Epithelial-derived neutrophil-activating peptide-78
Tumor necrosis factor-α receptor
Intercellular adhesion molecule-1
Triggering receptor expressed on myeloid cells -1
Phospholipase
Procalcitonin
Neopterin
Hydrogen sulphide

Conclusions

Traditional severity scores have been widely used by most physicians to detect severe AP. However, the complexity of the Ranson and APACHE II scales has prompted some authors to evaluate inflammatory mediators (summarized in table 1) as potential prognostic markers in AP. These markers can be proteases released from the pancreas or inflammatory proteins induced by the inflammation in the pancreas and distant organs. For most of them, the higher the blood concentration of these markers, the more severe the disease will be. IL-6, IL-8 and neutrophil elastase are the earliest factors to be detected and are readily measured with commercial kits.

However, despite the studies of predictors up to now, we are still lacking a unique and simple biological or physiological parameter that is easily measurable which we can safely rely on to predict the severity of an attack of AP.

The ideal predictive factor will show the ability to detect severe forms of the disease before the occurrence of multiple organ dysfunction. It is then conceivable that the therapeutic antagonism of these factors might prevent or attenuate the severity of the multiple organ dysfunction and, consequently, ameliorate the disease outcome.

References

1 Frossard J, Steer M, Pastor C: Acute pancreatitis. Lancet 2008;371:143–152.
2 Bhatia M, Wong FL, Cao Y, Lau HY, Huang J, Puneet P, Chevali L: Pathophysiology of acute pancreatitis. Pancreatology 2005;5:132–144.
3 Lund H, Tonnesen H, Tonnesen MH, Olsen O: Long-term recurrence and death rates after acute pancreatitis. Scand J Gastroenterol 2006;41:234–238.
4 Imrie CW: Acute pancreatitis: Overview. Eur J Gastroenterol Hepatol 1997;9:103–105.
5 Trapnell JE, Duncan EHL: Patterns of incidence in acute pancreatitis. Br Med J 1975;2:179–183.

6 Giggs JA, Bourke JB, Katschinski B: The epidemiology of primary acute pancreatitis in Greater Nottingham: 1969–1983. Soc Sci Med 1988;26:79–89.
7 Go VLW: Etiology and epidemiology of pancreatitis in the United States; in Bradley E (ed): Acute Pancreatitis: Diagnosis and Therapy. New York, Raven Press, 1994, pp 235–239.
8 Jaakkola M, Nordback I: Pancreatitis in Finland between 1970 and 1989. Gut 1993;34:1255–1260.
9 Renner IG, Savage WT, Pantoja JL, Renner VJ: Death due to acute pancreatitis: a retrospective analysis of 405 autopsy cases. Dig Dis Sci 1985;30:1005–1018.
10 Lankisch PG, Burchard-Reckert S, Petersen M, et al: Morbidity and mortality in 602 patients with acute pancreatitis seen between the years 1980–1994. Z Gastroenterol 1996;34:371–377.
11 Lankish P: Epidemiology of acute pancreatitis; in Buchler MW, Uhl W, Friess H, Malfertheiner P (eds): Acute Pancreatitis: Novel Concepts in Biology and Therapy. Berlin, Blackwell Science, 1999, pp 145–153.
12 Saluja AK, Steer M: Pathophysiology of pancreatitis: role of cytokines and other mediators of inflammation. Digestion 1999;60:27–33.
13 Norman JG, Fink GW, Messina J, Carter G, Franz MG: Timing of tumor necrosis factor antagonism is critical in determining outcome in murine lethal acute pancreatitis. Surgery 1996;120:515–521.
14 Gloor B, Todd KE, Lane JS, Rigberg DA, Reber HA: Mechanism of increased lung injury after acute pancreatitis in IL-10 knockout mice. J Surg Res 1998;80:110–114.
15 Bhatia M, Saluja AK, Hofbauer B, Lee H-S, Frossard J-L, Steer ML: The effects of neutrophil depletion on a completely non-invasive model of acute pancreatitis-associated lung injury. Int J Pancreatol 1998;24:77–83.
16 Bhatia M, Brady M, Kang YK, Costello E, Newton DJ, Christmas S, Neoptolemos JP, Slavin J: MCP-1 but not CINC synthesis is increased in rat pancreatic acini in response to caerulein hyperstimulation. Am J Physiol Gastrointest Liver Physiol 2002;282:G77–G85.
17 He M, Horuk R, Bhatia M: Treatment with BX471, a non-peptide CCR1 antagonist, protects mice against acute pancreatitis-associated lung injury by modulating neutrophil recruitment. Pancreas 2007;34:233–241.
18 Frossard JL, Saluja A, Bhagat L, et al: The role of intracellular adhesion molecule 1 and neutrophils in acute pancreatitis and pancreatitis-associated lung injury. Gastroenterology 1999;116:694–701.
19 Poch B, Gansauge F, Rau B, et al: The role of polymorphonuclear leukocytes and oxygen-derived free radicals in experimental acute pancreatitis: mediators of local destruction and activators of inflammation. FEBS Lett 1999;19:268–272.
20 Pastor CM, Matthay M, Frossard JL: Pancreatitis-associated lung injury: new insights. Chest 2003;124:2341–2351.
21 Bhatia M, Saluja AK, Hofbauer B, Lee H-S, Frossard J-L, Castagliuolo I, Wang C-C, Gerard N, Pothoulakis C, Steer ML: Role of NK1 receptor in the development of acute pancreatitis and pancreatitis-associated lung injury. Proc Natl Acad Sci USA 1998;95:4760–4765.
22 Lau HY, Wong FL, Bhatia M: A key role of neurokinin 1 receptors in acute pancreatitis and associated lung injury. Biochem Biophys Res Commun 2005;327:509–515.
23 Bhatia M, Wong FL, Fu D, Lau HY, Moochhala S, Moore PK: Role of hydrogen sulfide in acute pancreatitis and associated lung injury. FASEB J 2005;19:623–625.
24 Bhatia M, Sidhapuriwala J, Ng SW, Tamizhselvi R, Moochhala S: Proinflammatory effects of hydrogen sulfide on substance P in caerulein-induced acute pancreatitis. J Cell Mol Med 2008;12:580–590.
25 Ranson JH, Rifkind KM, Turner JW: Prognostic signs and nonoperative peritoneal lavage in acute pancreatitis. Surg Gynecol Obstet 1976 143:209–219.
26 Ranson JH, Pasternack BS: Statistical methods for quantifying the severity of clinical acute pancreatitis. J Surg Res 1977;22:79–91.
27 Blamey SL, Imrie CW, J ON, Gilmour WH, Carter DC: Prognostic factors in acute pancreatitis. Gut 1984;25:1340–1346.
28 Wilson C, Heath DI, Imrie CW: Prediction of outcome in acute pancreatitis: a comparative study of APACHE II, clinical assessment and multiple factor scoring systems. Br J Surg 1990;77:1260–1264.
29 Robert JH, Frossard JL, Mermillod B, et al: Early prediction in acute pancreatitis: prospective study comparing CT scan, Ranson, Glasgow, APACHE II scores and various serum markers. World J Surg 2002;26:612–619.
30 Frossard JL: TAP: from pathophysiology to clinical usefulness. JOP 2001;2:69–77.
31 Frossard JL, Hadengue A, Pastor CM: New serum markers for the detection of severe acute pancreatitis in humans. Am J Respir Crit Care Med 2001;164:162–170.
32 Buchler M: Objectification of the severity of acute pancreatitis. Hepatogastroenterology 1991;38:101–108.
33 Wilson C, Heads A, Shenkin A, Imrie CW: C-reactive protein, antiproteases and complement factors as objective markers of severity in acute pancreatitis. Br J Surg 1989;76:177–181.

34 Leese T, Shaw D, Holliday M: Prognostic markers in acute pancreatitis: can pancreatic necrosis be predicted? Ann R Coll Surg Engl 1988;70:227–232.

35 Gross V, Scholmerich J, Leser HG, et al: Granulocyte elastase in assessment of severity of acute pancreatitis: comparison with acute-phase proteins C-reactive protein, alpha 1- antitrypsin, and protease inhibitor alpha 2-macroglobulin. Dig Dis Sci 1990;35:97–105.

36 Duarte-Rojo A, Suazo-Barahona J, Ramirez-Iglesias M, Uscanga L, Robles-Díaz G: Time frames for analysis of inflammatory mediators in acute pancreatitis: improving admission triage. Dig Dis Sci 2008, E-pub ahead of print.

37 Paajanen H, Laato M, Jaakkola M, Pulkki K, Niinikoski J, Nordback I: Serum tumour necrosis factor compared with C-reactive protein in the early assessment of severity of acute pancreatitis. Br J Surg 1995;82:271–273.

38 Chen C, Wang S, Lee F, Chang F, Lee S: Proinflammatory cytokines in early assessment of the prognosis of acute pancreatitis. Am J Gastroenterol 1999;94:213–218.

39 McKay C, Gallagher G, Brooks B, Imrie C, Baxter J: Increased monocyte cytokine production in association with systemic complications in acute pancreatitis. Br J Surg 1996;83:919–923.

40 Brivet E, Emilie G, Galanaud P: Pro- and anti-inflammatory cytokines during acute severe pancreatitis: an early and sustained response, although unpredictable of death: Parisian Study Group on Acute Pancreatitis. Crit Care Med 1999;27:749–755.

41 Matsuda K, Hirasawa H, Oda S: Current topics on cytokine removal technologies. Ther Apher 2001;5:306–314.

42 Leser HG, Gross V, Scheibenbogen C, et al: Elevation of serum interleukin-6 concentration precedes acute-phase response and reflects severity in acute pancreatitis. Gastroenterology 1991;101:782–785.

43 Heath DI, Cruickshank A, Gudgeon M, Jehanli A, Shenkin A, Imrie CW: Role of interleukin-6 in mediating the acute phase protein response and potential as an early means of severity assessment in acute pancreatitis. Gut 1993;34:41–45.

44 de-Madaria E, Martínez J, Sempere L, et al: Cytokine genotypes in acute pancreatitis: association with etiology, severity, and cytokine levels in blood. Pancreas 2008;37:295–301.

45 Watanabe E, Hirasawa H, Oda S, Matsuda K, Hatano M, Tokuhisa T: Extremely high interleukin-6 blood levels and outcome in the critically ill are associated with tumor necrosis factor – and interleukin-1 – related gene polymorphisms. j 2005;33:89–97.

46 Gross V, Andreesen R, Leser H, et al: Interleukin-8 and neutrophil activation in acute pancreatitis. Eur J Clin Invest 1992;22:200–203.

47 Shokuhi S, Bhatia M, Christmas, S, Neoptolemos, JP, Slavin J: ENA78 and Gro-α are predictors of the severity of clinical acute pancreatitis. Br J Surg 2002; 89:566–572.

48 Hofner P, Baloq A, Gyulai Z, et al: Polymorphism in the IL-8 gene, but not in the TLR4 gene, increases the severity of acute pancreatitis. Pancreatology 2006;6:542–548.

49 Dominguez-Munoz JE, Carballo F, Garcia MJ, et al: Clinical usefulness of polymorphonuclear elastase in predicting the severity of acute pancreatitis: results of a multicentre study. Br J Surg 1991;78: 1230–1234.

50 Uhl W, Buchler M, Malfertheiner P, Martini M, Beger HG: PMN-elastase in comparison with CRP, antiproteases, and LDH as indicators of necrosis in human acute pancreatitis. Pancreas 1991;6:253–259.

51 Van Laethem J, Marchant A, Delvaux A, et al: Interleukin 10 prevents necrosis in murine experimental acute pancreatitis. Gastroenterology 1995; 108:1917–1922.

52 Rongione AJ, Kusske AM, Kwan K, Ashley SW, Reber HA, McFadden DW: Interleukin 10 reduces the severity of acute pancreatitis in rats. Gastroenterology 1997;112:960–967.

53 Pezzilli R, Billi P, Miniero R, Barakat B: Serum interleukin-10 in human acute pancreatitis. Dig Dis Sci 1997;42:1469–1472.

54 Chen CC, Wang SS, Lu RH, Chang FY, Lee SD: Serum interleukin 10 and interleukin 11 in patients with acute pancreatitis. Gut 1999;45:895–899.

55 Stimac D, Fisic E, Milic S, Bilic-Zulle L, Peric R: Prognostic values of IL-6, IL-8, and IL-10 in acute pancreatitis. J Clin Gastroenterol 2007;40:209–212.

56 Kylänpää-Bäck M, Takala A, Kemppainen E, et al: Procalcitonin, soluble interleukin-2 receptor, and soluble E-selectin in predicting the severity of acute pancreatitis. Crit Care Med 2001;29:63–69.

57 Pezzilli R, Miniero R, Cappelletti O, Barakat B: Behavior of serum interleukin 12 in human acute pancreatitis. Pancreas 1999;18:247–251.

58 Ueda T, Takeyama Y, Yasuda T, et al: Serum interleukin-15 level is a useful predictor of the complications and mortality in severe acute pancreatitis. Surgery 2007;142:319–326.

59 Ueda T, Takeyama Y, Yasuda T, et al: Significant elevation of serum interleukin-18 levels in patients with acute pancreatitis. J Gastroenterol 2006;41:158–165.

60 Kaufmann P, Tilz GP, Lueger A, Demel U: Elevated plasma levels of soluble tumor necrosis factor receptor (sTNFRp60) reflect severity of acute pancreatitis. Intensive Care Med 1997;23:841–848.

61 Kaufmann P, Tilz GP, Smolle KH, Demel U, Krejs GJ: Increased plasma concentrations of circulating intercellular adhesion molecule-1 (cICAM-1) in patients with necrotizing pancreatitis. Immunobiology 1996;195:209–219.
62 Bouchon A, Dietrich J, Colonna M: Cutting edge: inflammatory responses can be triggered by TREM-1, a novel receptor expressed on neutrophils and monocytes. J Immunol 2000;164:4991–4995.
63 Yasuda T, Takeyama Y, Ueda T, et al: Increased levels of soluble triggering receptor expressed on myeloid cells-1 in patients with acute pancreatitis. Crit Care Med 2008;36:2048–2053.
64 Viedma JA, Perez-Mateo M, Dominguez JE, Carballo F: Role of interleukin-6 in acute pancreatitis: comparison with C-reactive protein and phospholipase A. Gut 1992;33:1264–1267.
65 Uomo G, Spada OA, Manes G, et al: Neopterin in acute pancreatitis. Scand J Gastroenterol 1996;31: 1032–1036.
66 Mora A, Perez-Mateo M, Viedma JA, Carballo F, Sanchez-Paya J, Liras G: Activation of cellular immune response in acute pancreatitis. Gut 1997; 40:794–797.
67 Kaufmann P, Tilz GP, Demel U, Wachter H, Kreijs GJ, Fuchs D: Neopterin plasma concentrations predict the course of severe acute pancreatitis. Clin Chem Lab Med 1998;36:29–34.
68 Fernandes ML, Ho KY, Wee EW, Ng SW, He M, Bhatia M: Serum hydrogen sulfide: a novel biological marker that predicts severity of acute pancreatitis at its early stage. Pancreatology 2008;8:326.
69 Knaus WA, Draper EA, Wagner DP, Zimmerman JE: APACHE II: a severity of disease classification system. Crit Care Med 1985;12:818–829.

Prof. Jean Louis Frossard, MD
Division of Gastroenterology, Geneva University Hospitals
CH–1211 Genève 14 (Switzerland)
Tel. +22 372 93 40, Fax +22 372 93 66, E-Mail jean-louis.frossard@hcuge.ch

Mayerle J, Tilg H (eds): Clinical Update on Inflammatory Disorders of the Gastrointestinal Tract.
Front Gastrointest Res. Basel, Karger, 2010, vol 26, pp 157–165

Antibiotics, Probiotics and Enteral Nutrition: Means to Prevent Infected Necrosis in AP

Ingrid A. van Doesburg · Marc G. Besselink · Olaf J. Bakker · Hjalmar C. van Santvoort · Hein G. Gooszen, on behalf of the Dutch Pancreatitis Study Group

Department of Surgery, University Medical Center Utrecht, Utrecht, The Netherlands

Abstract

Mortality in severe acute pancreatitis (AP) is predominantly associated with secondary infection of (peri-)pancreatic necrosis and other infectious complications. Suggested prophylactic strategies are antibiotics, probiotics and enteral nutrition. We performed a literature review on the prophylactic treatment with antibiotics, probiotics and enteral nutrition in AP. The 2 most recent double blind placebo-controlled randomized trials and 2 meta-analyses on systemic antibiotic prophylaxis in AP did not demonstrate a significant reduction of infected (peri-)pancreatic necrosis. The earlier findings that probiotics are effective in preventing infections in AP were not confirmed by a recent multicenter placebo-controlled randomized controlled trial. Unexpectedly, mortality was significantly higher in patients receiving probiotics. Enteral nutrition, when compared to parenteral nutrition, reduces infectious complications and mortality in AP. The magnitude of the effect may depend on timing of the start of enteral feeding. Current evidence does not support the use of prophylactic antibiotics or probiotics in patients with severe AP. Enteral nutrition should be preferred over parenteral nutrition. Optimal timing of commencement of enteral nutrition should be investigated in randomized studies.

Introduction

Approximately one fifth of all patients diagnosed with AP will develop severe AP [1]. Severe AP is associated with an overall mortality rate up to 30% [1, 2].

Mortality in severe AP has a bimodal distribution during the course of the disease. In the first 1–2 weeks after onset of symptoms, 'early mortality' is attributed to a systemic inflammatory response syndrome, if associated with persistent (multiple) organ failure. After surviving this episode, 'late mortality' is mainly caused by sepsis and multiple organ failure due to secondary infections such as infected pancreatic necrosis, bacteraemia and pneumonia [3]. Infectious complications are associated with about 80% of mortality in severe AP [1, 4].

One of the first steps in the process of secondary infection in severe AP is bacterial translocation: the phenomenon that enteral bacteria cross the gastrointestinal mucosal barrier and invade the systemic compartment [5]. In experimental and clinical studies, bacterial translocation is believed to be the result of a cascade of events depending on a disturbance of host-bacterial interactions on 3 levels: (1) the intestinal lumen – impaired small bowel motility and bacterial overgrowth, (2) the intestinal epithelium – structural mucosal barrier failure leading to increased gut permeability [6], and (3) the systemic immune system – a dysregulation of the pro- and anti-inflammatory balance [7]. A recent experimental study in rats suggested that mesenteric lymphogene transmission is also involved in the process of bacterial translocation [8].

The main treatment strategies that have been suggested to prevent infection of (peri-)pancreatic necrosis and other infectious complications in AP are prophylactic administration of antibiotics, probiotics and enteral nutrition (EN). These treatment strategies will be consecutively discussed.

Antibiotics

The prophylactic use of systemic antibiotics in AP has been extensively studied over the last decades. The rationale for prophylactic treatment is to diminish potential haematogenous spread of pathogens after bacterial translocation has occurred.

The efficacy of various types of antibiotics in relation to AP has been studied in vitro and in experimental studies. The agents with greatest tissue penetration and bactericidal properties are carbapenems, fluoroquinolones, metronidazoles and cephalosporines. Imipenem especially has a relevant antimicrobial spectrum and effective penetration of (peri-)pancreatic necrosis and is also considered as an immunomodulatory agent [9].

Several randomized controlled trials (RCTs) studied the effect of systemic antibiotic prophylaxis on prevention of infection of pancreatic necrosis [10, 11]. In all these studies, the number of included patients was small, there was no placebo control, methodology was debatable and results were inconsistent. The results of the various meta-analyses showed inconsistency as well, which is probably explained by the fact that different combinations of trials have been pooled. The 2 most recent RCTs on this topic, by Isenmann et al. [12] and Dellinger et al. [13], were double-blind placebo-controlled trials and are considered to be of high methodological quality. Neither trial demonstrated beneficial effects of antibiotics in reducing (peri-)pancreatic infections, extra pancreatic infections (e.g. bacteraemia, pneumonia) or mortality. Notably, one of the studies demonstrated a significant increase in infections with bacteria resistant to the type of antibiotics administered [12]. The most recently updated meta-analyses underscored the negative results of the last 2 RCTs [14, 15]. Interestingly, a significant relationship was demonstrated between the (high) methodological quality of these RCTs and the (small) effect on mortality [14].

The total numbers of patients included in the 2 double-blind placebo-controlled trials were relatively small (114 and 100 patients respectively). It cannot be ruled out that antibiotics might be effective in reducing infections in AP, but if the effect exists it is probably quite small.

The conflicting results of the various studies have led to confronting opinions on the relative benefits and disabilities of prophylactic antibiotics. If antibiotic use does not benefit the patient in terms of prevention of infected necrosis and mortality, this raises the question of whether their use should be avoided. Broad-spectrum antibiotic prophylaxis may lead to bacterial resistance [16], fungal infections [17], selective overgrowth [18] of pathogens and increased costs [19]. Nowadays, most pancreatologists refrain from using antibiotic prophylaxis in patients with (severe) AP.

Currently in many intensive care units, patients are prophylactically treated by selective digestive tract decontamination (SDD). The rationale for SDD is different from that of prophylactic systemic antibiotics. SDD aims to prevent bacterial translocation from the gut prior to infection. In recent studies with general patient populations, no subgroup analysis for patients with AP has been performed. The only RCT of SDD in patients with severe AP demonstrated a significant reduction of Gram-negative bacterial colonization of the digestive tract, and a significant reduction of morbidity and mortality [20]. Due to the moderate methodological quality (an underpowered study with lack of various definitions and indications and which was not double blinded) and the overall scarceness of evidence in severe AP, SDD is not considered standard practice in severe AP.

Conclusion

Current evidence does not support routine antibiotic prophylaxis in patients with severe AP.

Probiotics

The World Health Organization defines probiotics as follows: 'Probiotics are live micro-organisms which when administered in adequate amounts confer a health benefit on the host' [21]. Probiotics can be administered together with prebiotics (synbiotics), non-digestible fibre supplements that, in addition to probiotics, enhance its activity.

Probiotics have been suggested to reduce bacterial translocation (in AP) through a beneficial effect on the disturbances at the levels of host-bacterial interactions; the intestinal lumen, the intestinal epithelium and the immune system.

In the intestinal lumen, selected probiotic strains may prevent bacterial overgrowth of potential pathogens by a direct antimicrobial effect and competitive growth

[22]. At the level of the intestinal epithelium, probiotics may preserve or reinforce the mucosal gastrointestinal barrier function. This is achieved through several mechanisms: prevention of bacterial adherence to the epithelium by competitive exclusion, inhibition of pathogen-induced increase of epithelial permeability [6] and regulation of enterocyte gene expression involved in the maintenance of the mucosal barrier [23]. Selected probiotic strains have been found to be capable of inhibiting local pro-inflammatory reactions in enterocytes after stimuli such as pathogenic bacterial adhesion or ischaemia [23]. Finally, probiotics are thought to have a regulatory effect on the mucosal and systemic immune system; in vitro selected probiotic strains induce production of the anti-inflammatory cytokine IL-10 by monocytes and lymphocytes [24].

Various experimental studies on the prophylactic role of probiotics in AP have been performed. In an experimental study in rats where pancreatitis was induced, probiotics reduced bacterial overgrowth of potential pathogens in the duodenum, resulting in reduced bacterial translocation to extra-intestinal sites [25]. This reduction was associated with a decrease of morbidity and late mortality.

In the clinical setting, prevention of infectious complications with prophylactic probiotics has been studied in several RCTs. In patients undergoing major abdominal surgery (liver transplantation, liver resection, pancreaticoduodenectomy) the administration of pre- and proboitics significantly reduced the incidence of post-operative infections, although operative complications and post-operative infections have been lumped together in some of these studies [26–28].

Notably, methodologic quality, as well as several other factors (e.g. variation in dose and type of probiotic species) varied greatly between studies, which may have influenced the results.

Three RCTs have studied the role of prophylactic probiotics in AP. Olah et al. [29] performed the first double-blind placebo-controlled trial in 45 patients suffering from AP. Probiotics *(Lactobacillus plantarum)* were found to significantly reduce the incidence of infected pancreatic necrosis. The same group performed a second trial and randomized 62 patients with predicted severe AP between treatment with prebiotics combined with probiotics (Symbiotic 2000) and prebiotics only [30]. No effect of probiotics on the incidence of multiple organ failure, septic complications or mortality was detected. Only the combined incidence of systemic inflammatory response syndrome and multiple organ failure was significantly reduced in the probiotics group.

The most recent and largest randomized, double-blind, placebo-controlled trial on probiotics in AP (PROPATRIA; probiotics in pancreatitis trial) was performed by the Dutch Pancreatitis Study Group [31]. This multicentre study randomized 296 patients with predicted severe AP to a multispecies probiotic preparation (Ecologic 641) or placebo.

Probiotic prophylaxis did not reduce the rate of infectious complications. Unexpectedly, mortality was twice as high in patients treated with probiotics. The results suggested that bowel ischaemia contributed to the higher mortality, although the exact mechanism for this adverse event is currently unclear.

Subsequently, the effect of the combination of probiotic strains used on the relationship between intestinal barrier function, bacterial translocation and clinical outcome was studied in a consecutive subset of the patients randomized in the trial [32]. In patients with concomitant organ failure that received probiotics, markers of enterocyte damage and bacterial translocation in the urine were increased, but markers for intestinal permeability were not altered. In patients without organ failure, probiotic prophylaxis was actually associated with a reduction in a marker for bacterial translocation. Bacteraemia, infected necrosis, organ failure and mortality were all associated with intestinal barrier dysfunction early in the course of AP.

How to explain this deleterious effect of probiotics in patients with organ failure? It is known that the bloodflow of the intestinal mucosa is reduced by up to 85% in rats suffering from severe AP [33]. If a similar phenomenon is assumed in patients with AP, we may hypothesize that administration of the probiotic bacteria on top of enteral feeding might increase oxygen demand at the mucosal level, with a further deleterious effect on the already compromised intestinal barrier. Another possible explanation may be that, in the subgroup of most severely ill patients, the probiotics caused local inflammation at the mucosal level leading to decreased oxygen supply, and, ultimately, ischaemia.

Conclusion

Based on the outcome of the recent Dutch PROPATRIA trial, the use of probiotics must be advised against in patients with severe AP and in general in critically ill patients needing circulatory and ventilatory support in the intensive care unit.

Research on probiotic prophylaxis in patients with organ failure has been set back to the experimental stage to study the possible mechanism of adverse events recently observed in PROPATRIA.

Enteral Nutrition

In the previous century, patients with predicted severe AP were kept on a nil by mouth regimen because enteral nutrient intake was considered harmful. Nutrients passing the duodenum were thought to stimulate pancreatic enzyme secretion, leading to premature activation of proteolytic enzymes within the acinar cells of the pancreas, subsequently increasing tissue injury and aggravating the disease course [34]. Parenteral nutrition (PN) was considered to supply essential nutrients without stimulating the pancreas and was therefore considered superior in patients with AP.

Fasting, however, causes atrophy of the enteral mucosa and decreases intestinal motility [35]. Consequently, small bowel bacterial overgrowth occurs and this, together with an increase in intestinal permeability, may attribute to bacterial translocation

[36]. Given the above, there is no theoretical benefit of 'keeping the pancreas at rest'. Moreover, clinical evidence for this strategy is scarce [34].

During the last decades, several RCT's comparing EN with PN demonstrated a significant reduction in the number of septic complications with use of EN [37, 38]. These studies, however, all included relatively few patients and differed considerably in eligibility criteria. A more recent RCT showed that EN, as compared to PN, significantly reduced (peri-)pancreatic necrosis, multiple organ failure and mortality [39].

The supposed main mechanism for the superiority of EN over PN is the maintenance of the intestinal mucosal barrier function through suppletion of intraluminal nutrition to enterocytes with subsequent reduction of both bacterial overgrowth and bacterial translocation [40].

In addition to prevention of infectious complications, EN also appears to be superior to PN in avoidance of indwelling catheter infections, improving tolerance of oral refeeding but also in cost-effectiveness and length of stay [37, 38]. A practical limitation of EN is that some patients do not tolerate the mechanical discomfort of a nasojejunal tube.

In most centres, EN is administered through a nasojejunal tube. The rationale for using the nasojejunal route is, again, to avoid nutrients passing the duodenum inducing stimulation of pancreatic enzyme secretion and to diminish the risk of pulmonary aspiration of gastric contents.

Few studies have compared the nasojejunal (technically more demanding and more expensive) with the nasogastric feeding tube route. Two recent RCTs did not show any difference in safety, morbidity or mortality in severe AP between nasojejunal and nasogastric feeding, nor was an aggravation of the pancreatic inflammation demonstrated [41, 42]. Because of a small size of these studies, more studies are needed to establish whether nasogastric feeding is truly equal to nasojejunal feeding.

There is no consensus on the optimal timing of the start of enteral feeding in AP [37, 43]. Theoretically, starting enteral feeding as soon as possible after onset of symptoms will contribute to maintaining the intestinal barrier function and thereby mitigate the clinical course. A very early start of EN has been shown to be beneficial in critically ill patients with diseases other than AP [44]. Recently, a systematic review including non-randomized clinical studies, demonstrated beneficial effects of early start of EN (24 or <48 h after admission) on multiple organ failure, pancreatic infectious complications and mortality in severe AP [44].

Conclusion

EN is superior to PN in AP. The aim of EN is not only to provide sufficient caloric intake but also to reduce (pancreatic) infectious complications and to promote faster recovery from the disease.

Because the impact of these effects on the clinical course may depend on timing of the start of EN, a RCT comparing early start of EN with delayed EN in patients with AP seems a logical next step.

Summary

In clinical practice, based on the evidence presented in this chapter, the role of prophylactic systemic antibiotics in patients with severe AP has, if ever present, become limited.

The authors advise against prophylactic probiotics in patients with severe AP, at least until the mechanism for the observed severe adverse events in patients with organ failure has been clarified in experimental studies.

The evidence that EN is superior to PN in the reduction of infected (peri-)pancreatic necrosis in patients with in severe AP has been well accepted. The potential influencing effect of timing of EN is currently subject of further studies.

References

1 Working party of the British Society of Gastroenterology; association of surgeons of Great Britain and Ireland; pancreatic society of Great Britain and Ireland; association of upper GI surgeons of Great Britain and Ireland: UK guidelines for the management of acute pancreatitis. Gut 2005;54(suppl 3): iii1–iii9.

2 Banks PA, Freeman ML: Practice Parameters Committee of the American College of Gastroenterology: Practice guidelines in acute pancreatitis. Am J Gastroenterol 2006;101:2379–2400.

3 Buter A, Imrie CW, Carter CR, Evans S, McKay CJ: Dynamic nature of early organ dysfunction determines outcome in acute pancreatitis. Br J Surg 2002; 89:298–302.

4 Besselink MG, van Santvoort HC, Boermeester MA, Nieuwenhuijs VB, van Goor H, Dejong CH, Schaapherder AF, Gooszen HG; Dutch Acute Pancreatitis Study Group: Timing and impact of infections in acute pancreatitis. Br J Surg 2009;96: 267–273

5 Guarner F: Gut flora in health and disease. Lancet 2003;361:512–519

6 Lutgendorff F, Nijmeijer RM, Sandström PA, Trulsson LM, Magnusson KE, Timmerman HM, van Minnen LP, Rijkers GT, Gooszen HG, Akkermans LM, Söderholm JD. Probiotics prevent intestinal barrier dysfunction in acute pancreatitis in rats via induction of ileal mucosal glutathione biosynthesis. PLoS ONE 2009;4:e 4512.

7 Ammori BJ, Leeder PC, King RF, Barclay GR, Martin IG, Larvin M, McMahon MJ: Early increase in intestinal permeability in patients with severe acute pancreatitis: correlation with endotoxemia, organ failure, and mortality. J Gastrointest Surg 1999;3:252–262.

8 Mittal A, Phillips AR, Middleditch M, Ruggiero K, Loveday B, Delahunt B, Cooper GJ, Windsor JA: The proteome of mesenteric lymph during acute pancreatitis and implications for treatment. JOP 2009;10: 130–42

9 Buchler M, Malfertheiner P, Friess H, Isenmann R, Vanek E, Grimm, Schlegel P, Friess T, Beger HG: Human pancreatic tissue concentration of bactericidal antibiotics. Gastroenterology 1992;103:1902–1908.

10 Sainio V, Kemppainen E, Puolakkainen P, Taavitsainen M, Kivisaari L, Valtonen V, Haapiainen R, Schröder T, Kivilaakso E: Early antibiotic treatment in acute necrotising pancreatitis. Lancet 1995;346: 663–667.

11 Nordback I, Sand J, Saaristo R, Paajanen H: Early treatment with antibiotics reduces the need for surgery in acute necrotizing pancreatitis: a single-center randomized study. J Gastrointest Surg 2001;5: 113–118.

12 Isenmann R, Runzi M, Kron M, Kahl S, Kraus D, Jung N, Maier L, Malfertheiner P, Goebell H, Beger HG: Prophylactic antibiotic treatment in patients with predicted severe acute pancreatitis: a placebo-controlled, double-blinded trial. Gastroenterology 2004;126:997–1004.

13 Dellinger EP, Tellado JM, Soto NE, Ashley SW, Barie PS, Dugernier T, Imrie CW, Johnson CD, Knaebel HP, Laterre PF, Maravi-Poma E, Kissler JJ, Sanchez-Garcia M, Utzolino S: Early antibiotic treatment for severe acute necrotizing pancreatitis: a randomized, double-blind, placebo-controlled study. Ann Surg 2007;245:674–683.

14 de Vries AC, Besselink MG, Buskens E, Ridwan BU, Schipper M, van Erpecum KJ, Gooszen HG: Randomized controlled trials of antibiotic prophylaxis in severe acute pancreatitis: relationship between methodological quality and outcome. Pancreatology 2007;7:531–538.
15 Wittau M, Hohl K, Mayer J, Henne-Bruns D, Isenmann R: The weak evidence base for antibiotic prophylaxis in severe acute pancreatitis. Hepatogastroenterology 2008;55:2233–2237.
16 De Waele JJ, Vogelaers D, Hoste E, Blot S, Colardyn F: Emergence of antibiotic resistance in infected pancreatic necrosis. Arch Surg 2004;139:1371–1375.
17 Isenmann R, Schwarz M, Rau B, Trautmann M, Schober W, Beger HG: Characteristics of infection with *Candida* species in patients with necrotizing pancreatitis. World J Surg 2002;26:372–376.
18 Wren SM, Ahmed N, Jamal A, Safadi BY: Preoperative oral antibiotics in colorectal surgery increase the rate of *Clostridium difficile* colitis. Arch Surg 2005;140:752–756.
19 Berild D, Ringertz SH, Lelek M, Fosse B: Antibiotic guidelines lead to reductions in the use and cost of antibiotics in a university hospital. Scand J Infect Dis 2001;33:63–67.
20 Luiten EJ, Hop WC, Lange JF, Bruining HA: Controlled clinical trial of selective decontamination for the treatment of severe acute pancreatitis. Ann Surg 1995;222:57–65.
21 FAO/WHO: Health and Nutritional Properties of Probiotics in Food including Powder Milk with Live Lactic Acid Bacteria: Report of a Joint FAO/WHO Expert Consultation on Evaluation of Health and Nutritional Properties of Probiotics in Food Including Powder Milk with Live Lactic Acid Bacteria. Geneva, WHO, 2001.
22 Servin AL: Antagonistic activities of lactobacilli and bifidobacteria against microbial pathogens. FEMS Microbiol Rev 2004;28:405–440.
23 Marco ML, Pavan S, Kleerebezem M: Towards understanding molecular modes of probiotic action. Curr Opin Biotechnol 2006;17:204–210.
24 Niers LE, Timmerman HM, Rijkers GT, Van Bleek GM, van Uden NO, Knol EF, Kapsenberg ML, Kimpen JL, Hoekstra MO: Identification of strong interleukin-10 inducing lactic acid bacteria which down-regulate T helper type 2 cytokines. Clin Exp Allergy 2005;35:1481–1489.
25 van Minnen LP, Timmerman HM, Lutgendorff F, Verheem A, Harmsen W, Konstantinov SR, Smidt H, Visser MR, Rijkers GT, Gooszen HG, Akkermans LM: Modification of intestinal flora with multispecies probiotics reduces bacterial translocation and improves clinical course in a rat model of acute pancreatitis. Surgery 2007;141:470–480.
26 Nomura T, Tsuchiya Y, Nashimoto A, Yabusaki H, Takii Y, Nakagawa S, Sato N, Kanbayashi C, Tanaka O: Probiotics reduce infectious complications after pancreaticoduodenectomy. Hepatogastroenterology 2007;54:661–663.
27 Rayes N, Seehofer D, Theruvath T, Mogl M, Langreh JM, Nussler NC, Bengmark S, Neuhaus P: Effect of enteral nutrition and synbiotics on bacterial infection rates after pylorus-preserving pancreatoduodenectomy: a randomized, double blind trial. Ann Surg 2008;247:560–561.
28 Sugawara G, Nagino M, Nishio H, Ebata T, Takagi K, Asahara T, Nomoto K, Nimura Y: Perioperative synbiotic treatment to prevent postoperative infectious complications in biliary cancer surgery: a randomized controlled trial. Ann Surg 2006;244:706–714.
29 Olah A, Belagyi T, Issekutz A, Gamal ME, Bengmark S: Randomized clinical trial of specific lactobacillus and fibre supplement to early enteral nutrition in patients with acute pancreatitis. Br J Surg 2002;89: 1103–1107.
30 Oláh A, Belágyi T, Pótó L, Romics L Jr, Bengmark S: Synbiotic control of inflammation and infection in severe acute pancreatitis: a prospective, randomized, double blind study. Hepatogastroenterology 2007;54:590–594.
31 Besselink MG, van Santvoort HC, Buskens E, Boermeester MA, van Goor H, Timmerman HM, Nieuwenhuijs VB, Bollen TL, van Ramshorst B, Witteman BJ, Rosman C, Ploeg RJ, Brink MA, Schaapherder AF, Dejong CH, Wahab PJ, van Laarhoven CJ, van der Harst E, van Eijck CH, Cuesta MA, Akkermans LM, Gooszen HG; Dutch Acute Pancreatitis Study Group: Probiotic prophylaxis in predicted severe acute pancreatitis: a randomised, double-blind, placebo-controlled trial. Lancet 2008; 23;371:651–659.
32 Besselink M, Van Santvoort HC, Renooij W, De Smet MB, Boermeester MA, Fischer K, Timmerman HM, Ahmed Ali U, Cirkel GA, Bollen TL, van Ramshorst B, Schaapherder AF, Witteman BJ, Ploeg RJ, van Goor H, van Laarhoven CJ, Tan AC, Brink MA, van der Harst E, Wahab PJ, van Eijk CH, Dejong CH, van Erpecum KJ, Akkermans LM, Gooszen HG: Intestinal barrier dysfunction in a randomized trial of probiotic prophylaxis in acute pancreatitis. Ann Surg 2009, in press.
33 Andersson R, Wang X, Ihse I: The influence of abdominal sepsis on acute pancreatitis in rats: a study on mortality, permeability, arterial pressure, and intestinal blood flow. Pancreas 1995;11:365–373.
34 O'Keefe SJ, Lee RB, Anderson FP, Gennings C, Abou-Assi S, Clore J, Heuman D, Chey W: Physiological effects of enteral and parenteral feeding on pancreaticobiliary secretion in humans. Am J Physiol Gastrointest Liver Physiol 2003;284:G27–G36.

35 Ammori BJ: Role of the gut in the course of severe acute pancreatitis. Pancreas 2003;26:122–129.
36 Van Felius ID, Akkermans LM, Bosscha K, Verheem A, Harmsen W, Visser MR, Gooszen HG: Interdigestive small bowel motility and duodenal bacterial overgrowth in experimental acute pancreatitis. Neurogastroenterol Motil 2003;15:267–276.
37 Abou-Assi S, Craig K, O'Keefe SJ: Hypocaloric jejunal feeding is better than total parenteral nutrition in acute pancreatitis: results of a randomized comparative study. Am J Gastroenterol 2002;97:2255–2262.
38 Oláh A, Pardavi G, Belágyi T, Nagy A, Issekutz A, Mohamed GE: Early nasojejunal feeding in acute pancreatitis is associated with a lower complication rate, Nutrition 2002;18:259–262.
39 Petrov MS, Kukosh MV, Emelyanov NV: A randomized controlled trial of enteral versus parenteral feeding in patients with predicted severe acute pancreatitis shows a significant reduction in mortality and in infected pancreatic complications with total enteral nutrition. Dig Surg 2006;23:336–345.
40 Nagpal K, Minocha VR, Agrawal V, Kapur S: Evaluation of intestinal mucosal permeability function in patients with acute pancreatitis. Am J Surg 2006;192:24–28.
41 Eatock FC, Chong P, Menezes N, Murray L, McKay CJ, Carter CR, Imrie CW: A randomized study of early nasogastric versus nasojejunal feeding in severe acute pancreatitis. Am J Gastroenterol 2005;100:432–439.
42 Kumar A, Singh N, Prakash S, Saraya A, Joshi YK: Early enteral nutrition in severe acute pancreatitis: a prospective randomized controlled trial comparing nasojejunal and nasogastric routes. J Clin Gastroenterol 2006;40:431–434.
43 Petrov MS, Pylypchuk RD, Uchugina AF: A systematic review on the timing of artificial nutrition in acute pancreatitis. Br J Nutr 2009;101:787–793.
44 Marik PE, Zaloga GP: Early enteral nutrition in acutely ill patients: a systematic review. Crit Care Med 2001;29:2264–2270.

Hein Gooszen, MD, PhD
Dutch Pancreatitis Study Group, Department of Surgery
University Medical Center Utrecht
PO Box 85500, HP G04.228, NL–3508 GA Utrecht (The Netherlands)
Tel. +31 88 755 8074, Fax +31 30 254 1944, E-Mail h.gooszen@umcutrecht.nl

Mayerle J, Tilg H (eds): Clinical Update on Inflammatory Disorders of the Gastrointestinal Tract.
Front Gastrointest Res. Basel, Karger, 2010, vol 26, pp 166–175

IKK/NF-κB/Rel in Acute Pancreatitis and Pancreatic Cancer: Torments of Tantalus

Hana Algül · Roland M. Schmid

II. Medizinische Klinik, Klinikum rechts der Isar, Technical University of Munich, Munich, Germany

Abstract

Pancreatic ductal adenocarcinoma (PDA) and severe acute pancreatitis (AP) are still unresolved problems in routine clinicalpractice. Although the biologies of these diseases are fundamentally different, both forms share the problem of high morbidity, mortality and lack of specific therapeutic options. Therapy in these diseases is mostly symptomatic or non-specific. Over the last few years a huge body of evidence has emerged supporting a fundamental role of the IKK/NF-kB/Rel signalling pathway in the pathophysiology of pancreatic ductal adenocarcinoma and acute pancreatitis. The therapeutic potential and benefit of targeting NF-κB in pancreatic disorders and the possible complications and pitfalls of such an approach, are explored in this review.

Pancreatic ductal adenocarcinoma (PDA) and severe acute pancreatitis (AP) are still unresolved problems in routine clinical practice. Although the biologies of these diseases are fundamentally different, both forms share the problem of high morbidity, mortality and lack of specific therapeutic options.

PDA is the fourth leading cause of cancer mortality in the Western hemisphere, with an estimated 37,170 new cases and nearly the same number of estimated deaths in the USA in 2007. It is characterized by a dismal prognosis and still high lethality. Despite current multimodal therapy, treatment outcomes are poor, with an overall survival rate of less than 5%. Even in those 10% of patients who are eligible for potentially curative surgical resection only 10–15% survive longer than 5 years [1]. Similarily, approximately 25% of patients with AP develop a severe disease course that leads to systemic inflammatory response syndrome and sequelae such as multiorgan dysfunction syndrome and acute respiratory distress syndrome, with mortality rates up to 50%. Therapy in these patients is mostly symptomatic requiring mechanical ventilation and/or dialysis. The identification of relevant signalling pathways in inflammatory and malignant disorders of the pancreas might give rise to the development of specific therapies, thereby improving the outcome of these diseases [2].

Activation of transcription factors is a fundamental process in transferring to the nucleus signals from outside the cell or from its cytoplasm. Among many characterized transcription factors the activation of nuclear factor κB (NF-κB) proteins has been shown to play a key role in innate and adaptive immune responses, cell proliferation, cell death and inflammation. It has become clear that aberrant regulation of NF-κB and the signalling pathways that control its activity are involved in oncogenesis and inflammatory disorders. This article discusses recent evidence supporting the involvement of NF-κB in oncogenesis and inflammation in the pancreas. The therapeutic potential and benefit of targeting NF-κB in pancreatic disorders and the possible complications and pitfalls of such an approach, are explored [3].

Transcription Factor NF-κB

NF-κB is a ubiquitous eukaryotic transcription factor that belongs to a family of mammalian proteins referred to as the family of NF-κB/Rel proteins. These proteins are regulated at several levels: transcription, translation and post-translational processing. The protein family comprises the subunits of NF-κB1 (p50 and its precursor p105) and NF-κB2 (p52 and its precursor p100) and the Rel proteins RelAp65, RelB, and c-Rel. All of these proteins contain a highly conserved homology domain within the N-terminal 300 amino acids, the Rel homolgy domain (RHD). This domain is responsible for dimerization, nuclear localization, DNA binding, and interactions with proteins of the IκB family (IκBα, IκBβ, IκBγ, IκBε, Bcl-3). In its active DNA-binding form, NF-κB is a heterogeneous collection of dimers, composed of various combinations of the members of the NF-κB/Rel family. p65:p50 heterodimers were the first forms of NF-κB to be identified and are most abundant in all cell types. Heterodimeric complexes containing Rel proteins activate transcription, whereas homodimeric complexes of NF-κB1 (p50) and NF-κB2 (p52) exert inhibitory effects. It appears that all NF-κB complexes are regulated in the same manner by interaction with inhibitory proteins, termed IκBs [reviewed in 4].

In unstimulated cells, NF-κB is retained in the cytoplasm through interaction with the inhibitory proteins, termed IκBs. All IκBs contain either 6 or 7 ankyrin repeats, mediating binding to the RHD and masking the nuclear localization signal of NF-κB. IκBs also play an important role in termination of NF-κB activation. Newly synthesized IκB enters the nucleus and binds NF-κB, thereby enhancing its dissociation from the DNA. The complex is re-exported to the cytoplasm by means of a nuclear export sequence present in IκBs.

Two NF-κB activation pathways exist. The first, the classical pathway, is normally triggered in response to microbial and viral infections or exposure to pro-inflammatory cytokines. Many stimuli activate NF-κB, mostly through the IκB kinase (IKK)-dependent phosphorylation and subsequent degradation of IκB proteins. The IKK complex consists of 2 highly homologous kinase subunits, IKKα and IKKβ, and a

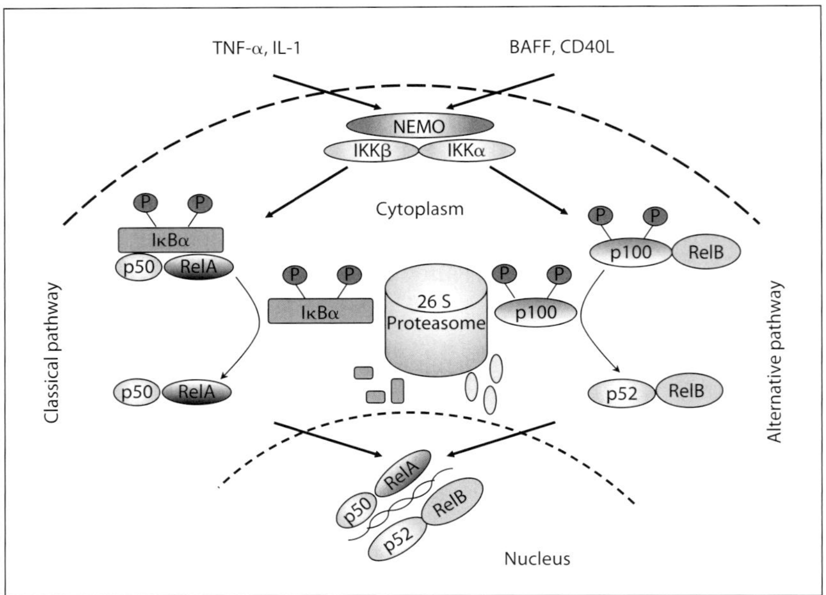

Fig. 1. The IKK/NF-κB/Rel signalling pathway. Different ligands induce the activation of the IKK complex consisting of the NEMO, IKKα and IKKβ subunits. The classical pathway involves the serine phosphorylation of the inhibitor protein IκBα which is degraded by the 26 S proteasoma complex. Heterodimers consisting of the p50:p65 complex translocate into the nucleus and bind to the DNA. The alternative pathway involves activation of IKKα and leads to the phosphorylation and processing of p100, generating p52:RelB heterodimers. Input signals for the alternative pathway follow ligation of LTβR, BAFFR, and CD40R. These alternative pathway stimuli also activate the classical pathway.

non-enzymatic regulatory component, IKKγ/NEMO. Phosphorylation of IκB rapidly occurs on 2 serine residues (Ser 32 and Ser 36 in human IκBα) by the IKK. The phosphorylation of IκB on serine residues targets it for the degradation by the 26 S proteasome. Once IκBα is degraded, the nuclear localization sequence of NF-κB is exposed, allowing translocation to the nucleus and activation of gene transcription. This pathway, which mostly targets p50:RelA and p50:c-Rel dimers, depends mainly on IKKβ activity. The other pathway, the alternative pathway, leads to selective activation of p52:RelB dimers by inducing processing of the NF-κB2/p100 precursor protein, which mostly occurs as a heterodimer with RelB in the cytoplasm. While both pathways regulate cell survival and death, the classical pathway mediates pro-inflammatory effects and is responsible for inhibition of programmed cell death under most conditions. The alternative pathway is important for survival of premature B cells and development of secondary lymphoid organs (fig. 1).

A series of knockout mouse models has been engineered illustrating the complex and essential function of NF-κB in the organism. The most impressive phenotype has been observed when RelA or IKKß were deleted in the whole organism. These mice undergo TNF-α dependent apoptosis in hepatocytes during embryonic development and result in prenatal lethality [reviewed in 5]. Due to the embryonic lethality of IKKβ and RelA deficient mice, analyses concerning the single organ-specific function of these proteins in several disease models have been hampered. The Cre-loxP strategy, however, has significantly contributed to identify the role of various members of the NF-κB/Rel protein family in different organs. This invaluable technology enables the organ-specific deletion or activation of genes of interest. Interestingly, both hepatocyte-specific deletion of IKK2 and RelA/p65 do not interfere with embryonic liver development further complicating the interpretation and relevance of previous studies. These mice do not show any liver pathologies [6]. However, the most valuable information concerning the role of the IKK/NF-κB/Rel signalling pathway has been described recently. Deletion of the IKKγ subunit in the colon, for example, demonstrate that defects in NF-κB signalling disrupts immune homeostasis in the gastrointestinal tract, causing an inflammatory-bowel-disease-like phenotype. Similarly, the hepatocytes specific deletion of NEMO unveils that NEMO-mediated NF-κB activation in hepatocytes plays an essential physiological role to prevent the spontaneous development of steatohepatitis and hepatocellular carcinoma, identifying NEMO as a tumour suppressor in the liver. In contrast a further mouse model focussing on the role of NF-κB in liver oncogenesis revealed anti-tumorous effects of IKKβ in vivo [reviewed in 7].

These recent data therefore shed a new light on the role of the IKK/NF-κB/Rel pathway in inflammation and carcinogenesis.

IKK/NF-κB/Rel Pathway in Acute Pancreatitis

Due to its involvement in chronic and acute inflammation the transcription factor NF-κB/Rel has become an interesting candidate for elucidating and validating therapeutic strategies in acute pancreatitis [8]. The first studies demonstrating the activation of NF-κB in acute pancreatitis were undertaken in cerulein-induced pancreatitis in rodents;however, subsequent studies have proven that NF-κB activation is also detectable in other AP models [9–14]. These findings suggest that activation of the IKK/NF-κB/Rel cascade marks a general, not model-restricted, pathway. Activation of the IKK complex and nuclear translocation of NF-κB occurs within 10 min after hyperstimulation with cerulein The activation cascade seems to follow a 2-phase time course. The initial increase in DNA-binding activity peaks at around 30 min and subsides by 1.5 h. At 3–6 h after cerulein infusion, a second wave of NF-κB/Rel can be detected [9, 11]. The inhibitor protein IκBα is rapidly but transiently degraded in pancreatitis [9, 11, 12]. At 3 and 6 h after cerulein infusion, IkBa returned to normal protein levels. Activation of NF-κB/Rel in acinar cells seems to depend on Ca^{2+}, calcineurin and protein kinase C activity and on specific stimuli [12, 13]. While the interference of IKK/NF-κB/Rel cascade with the conversion of trypsinogen and trypsin is still a matter of debate, the implication of NF-κB/Rel in cytokine and adhesion molecule gene expression, such as TNF-α, IL-6, Mob-1, MCP-1 and ICAM-1, and the implication of these substances in the immune and inflammatory responses is beyond doubt [14]. Presumably, upon release of these substances, mononuclear cells migrate to the site of inflammation and contribute to the second peak of NF-κB/Rel activation 3–6 h after the initiation of acute pancreatitis. Moreover, these hitherto unidentified soluble mediators seem to circulate in the bloodstream to activate NF-κB in macrophages and endothelial cells in distant organs, thereby perpetuating multiorgan inflammation. Therapeutic interventions to modulate NF-κB activation during AP, however, have been controversial so far. One approach to analyse the function of NF-κB during pancreatitis is to apply inhibitors in vivo models. Using non-specific chemicals, natural compounds, peptides or viral recombinant inhibitors several studies revealed attenuation of severity or even improved survival in different experimental models of AP (reviewed in [14]). Based on these findings it has been concluded that inhibition of this pathway might be a novel treatment approach for AP.

Using the CCK-induced model of AP in rats, one early study analysed the role of pharmacological NF-κB/Rel inhibitors (PDTC and NAC). Interestingly, inhibition of NF-κB resulted in perpetuated AP, an observation being diametrical to the previous reports [11]. From these data it was argued that the signalling cascade regulates protective genetic programmes and interferes with the fate of cell death towards necrosis or apoptosis. The difficulty with all of these studies is the use of inhibitors which are not totally specific and probably interfere with other signalling cascades.

Recent studies applying new genetically based techniques helped to shed light on this issue, but also challenged and complicated the current pathophysiological concept. In 2002, Chen et al. [15] reported that adenoviral mediated transfer of the active RelA/p65 subunit via intraductal retrograde injection into the pancreas was sufficient to provoke acute pancreatitis with lung inflammation in rats. Severity was reduced when IκBα was co-administered with RelA/p65. The difficulty with this study approach is the use of adenoviral infection. Adenoviral infection per se generates expression of chemokines and cytokines, and therefore the use of an inducible genetic system can be regarded as 'cleaner'. To overcome this pitfall, another study took advantage of the transgene technology in mice and revealed that ectopic overexpression of IKK2 (IKKβ) in the pancreas does not provoke AP, but worsens the course of cerulein pancreatitis [16]. Overexpression of IKK2 was achieved by an inducible system (rtTA/tetO) in mice which is driven by the CMV promoter. In response to doxycyclin the transgene was induced up to 200-fold, resulting in moderate activation of the IKK2 and nuclear translocation of RelA/p65. Although transgene expression was detected early on in the thymus and in the kidney, its expression seemed to be limited to the pancreas after 8–12 weeks. Interestingly, histological analysis of the pancreas revealed leucocyte infiltration without destruction of acinar cells as early as 3–4 weeks after the induction of the transgene. The patchy pattern of infiltrates were composed of B-lymphocytes and macrophages. Intra-acinar expression of TNF-α and RANTES were considered to be responsible for this observed phenotype. The authors therefore conclude that NF-κB activation per se in the pancreas does not lead to acute pancreatitis, but worsens the course of AP.

In a subsequent study of the same group, acinar-cell-restricted expression of IKK2 was achieved by using the rat elastase promoter [17]. This, surprisingly, was sufficient to induce local inflammation in the pancreas with acinar cell destruction, fibrosis and concomitant expression of proinflammatory cytokines. TNF-α was identified as important effector of IKK2-induced pancreatitis. Accordingly, acinar-cell-specific expression of a dominant-negative IKK2 abrogated the inflammatory phenotype and attenuated AP.

However, to overcome the problem with transgenic ectopic overexpression of the superordinated IKK2, another study adopted the Cre-loxP technology in mice to analyze the role of endogenous RelA/p65 in the exocrine pancreas during AP. Placing *loxP* sites between exons 7 and 10, the RHD of the *rela* gene was deleted by the homogenous acinar-cell-specific expression of the Cre-recombinase. RHD of the RelA/p65 protein accounts for the nuclear translocation of the heterodimers (p65/p50) and activation of the endogenous IKK/NF-κB/Rel pathway in the acinar cells is impaired. This mouse line develops normal pancreas, but, surprisingly, undergoes severe acute pancreatitis with increased lung and liver damage following cerulein stimulation [18]. As characterized RelA/p65 not only induces pro-inflammatory factors, but also regulates protective elements, such as the pancreatitis-associated protein I, which among other functions confers on acinar cells resistance to necrosis during inflammation [19].

In face of these diametrical results concerning the role of the IKK/NF-κB/Rel pathway, it remains difficult to make a conclusive statement. Of note, numerous studies have demonstrated that NF-κB is also activated outside of the pancreas during AP, which is not surprising given the fact that the inflammatory disease also affects extrapancreatic tissues, such as endothelial, myleoid cells, liver and lung [reviewed in 14]. Therefore, future efforts should aim to identify the specific role of the IKK/NF-κB/Rel signalling in these compartments by taking advantage of specific transgenic or conditional knock out mice (table 1). These strategies might help to clarify its role in AP and its relevance for therapeutic interventions.

IKK/NF-κB/Rel Pathway in Pancreatic Cancer

Histomorphological analyses of pancreatic cancer specimens from patients had long suggested that pancreatic ductal adenocarcinoma (PDA) does not develop de novo, but rather passes through a multistep progression, in which a series of non-invasive intraductal lesions of increasing histological severity eventually develop into an invasive cancer. Further histopathological findings have identified three percursor lesions: pancreatic intraepithelial neoplasia (PanIN), mucinous cystic neoplasm, and

Table 1. Genetic mouse strains to study the IKK/NF-κB/Rel pathway in the pancreas

System	Ref.
Cre-loxP System	
Conditional alleles	
ikkb$^{flox/flox}$	reviewed in [7]
ikk2-DK$^{flox/flox}$	reviewed in [7]
Nemo$^{flox/flox}$	reviewed in [7]
ikba$^{flox/flox}$	reviewed in [7]
rela$^{flox/flox}$	[18]
Pancreas-specific Cre expression	
Ptf1acre^{ex}	[18]
Pdx1-Cre	[33]
Elastase-Cre	[32]
Transgenic overexpression in the pancreas	
IKK2 constitutive active	[32]
IKK2 dominant negative	[32, 33]

intraductal papillary mucinous neoplasm. Among these lesions the PanINs which are found in the smaller-caliber pancreatic ducts have been extensively studied. PanINs are graded from stages I to III, with the earliest stage being characterized by the appearance of a columnar, mucinous eptihelium and with increasing architectural disorganization and nuclear atypia through stages II and III. Genetic studies revealed that activating *KRAS* mutations are the first genetic alterations, occurring sporadically in 30% of early neoplasms with the frequency rising to nearly 100% in advanced PDA. This gene encodes for a GTP-binding protein that regulates a wide variety of cellular functions including proliferation, differentiation, and survival. Subsequently, further genetic alterations (loss of function, amplifications, mutations) occur driving cells in the pancreas to malignant transformation. While the genetic repertoire has been in the focus, signalling pathways that drive the PanIN progressions have not been described so far [20].

Constitutive NF-κB activity is observed in many cancer, where it is thought to contribute to cell survival, angiogenesis, and invasion [21]. Given these qualities the IKK/NF-κB/Rel signalling pathway has become an interesting candidate in the pathophysiology of pancreatic cancer [22]. Two lines of evidence support a crucial role of this signalling pathway in pancreatic oncogenesis. The first line is based upon the observation that NF-κB is constitutively activated in 70% of human pancreatic cancers and in human pancreatic cell lines [23, 24]. This observation is not true for normal

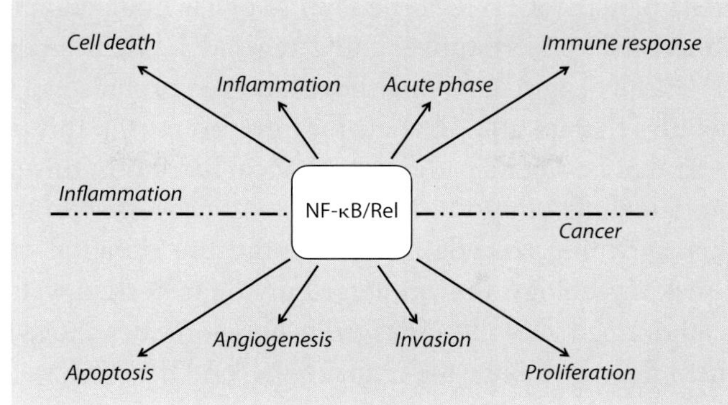

Fig. 2. Biological functions of NF-κB/Rel in inflammatory and malignant diseases of the pancreas.

pancreatic tissues or in immortalized, non-tumorigenic pancreatic epithelial cells. Interestingly, NF-κB activation has been shown to be crucial for RAS transformation of several cell types [25]. Expression of dominant negative Ras abrogates NF-κB activity in pancreatic cancer cell lines [26]. Inhibition of constitutive NF-κB activation in cells revealed the role of this signalling pathway in regulation of cell survival, VEGF, urokinase, and other pro-invasive or angiogenic factors [27, 28]. Moreover, interventions with this signalling cascade restored responsiveness of pancreatic cancer cell lines to various cytotoxic agents, probably via down-regulation of Bcl-2 and Bcl-XL and other anti-apoptotic proteins [29, 30]. Second, animal models for pancreatic tumours revealed NF-κB activation in pre-neoplastic lesions or classical PanIN-lesions as well as in advanced pancreatic carcinomas [31, 32]. Genetically engineered mice expressing mutated *KRAS* recapitulate the full spectrum of PanIN lesion and finally develop metastatic pancreatic cancer [33]. However, genetic or pharmacological evidence supporting a central role for this signalling pathway in PanIN progression or malignant transformation are still lacking. Therefore future efforts should be centered around the analysis of NF-κB activation in this setting using various NF-κB/Rel knockout mouse lines (table 1). These efforts might help to define a rationale for testing newly designed inhibitors of the IKK/NF-κB/Rel pathway in patients with pancreatic cancer in which currently a specific therapy is still not present but – due to the fatal outcome – urgently needed.

Conclusion

Due to its involvement in various biological processes, the transcription factor NF-κB/Rel has gained the interest of scientists working on inflammatory and malignant diseases of the pancreas. Evidence over the last few years has helped to validate this pathway as a therapeutic target (fig. 2). However, intervening in this pathway in acute pancreatitis, has turned out to be far more complicated than initially expected. Pancreatic cancer studies going beyond in vitro analysis or descriptive aspects in genetically engineered mice have not been provided so far, but are urgently needed to define a therapeutic rationale for interventions in this pathway. Finally, it has become clear that beside its pathophysiological functions, this pathway is implicated in other fundamentally important physiological processes. In this respect, interventions with this signalling cascade bring to mind the situation of Tantalus who, according to Greek mythology, was immersed up to his neck in water, but when he bent to drink, it all drained away; luscious fruits hung on trees above him, but when he reached for them the winds blew the branches beyond his reach.

References

1 Pliarchopoulou K, Pectasides D: Pancreatic cancer: current and future treatment strategies. Cancer Treat Rev 2009;35:431–436.

2 Frossard JL, Steer ML, Pastor CM: Acute pancreatitis. Lancet 2008;371:143–152.

3 Baud V, Karin M: Is NF-kappaB a good target for cancer therapy? Hopes and pitfalls. Nat Rev Drug Discov 2009;8:33–40.

4 Karin M: The IkappaB kinase: a bridge between inflammation and cancer. Cell Res 2008;18:334–342.

5 Sun B, Karin M: NF-kappaB signaling, liver disease and hepatoprotective agents. Oncogene 2008;27:6228–6244.

6 Geisler F, Algul H, Paxian S, Schmid RM: Genetic inactivation of RelA/p65 sensitizes adult mouse hepatocytes to TNF-induced apoptosis in vivo and in vitro. Gastroenterology 2007;132:2489–2503.

7 Pasparakis M, Luedde T, Schmidt-Supprian M: Dissection of the NF-kappaB signalling cascade in transgenic and knockout mice. Cell Death Differ 2006;13:861–872.

8 Algul H, Tando Y, Schneider G, Weidenbach H, Adler G, Schmid RM: Acute experimental pancreatitis and NF-kappaB/Rel activation. Pancreatology 2002;2:503–509.

9 Gukovsky I, Gukovskaya AS, Blinman TA, Zaninovic V, Pandol SJ: Early NF-kappaB activation is associated with hormone-induced pancreatitis. Am J Physiol 1998;275:G1402–G1414.

10 Satoh A, Shimosegawa T, Fujita M, Kimura K, Masamune A, Koizumi M, Toyota T: Inhibition of nuclear factor-kappaB activation improves the survival of rats with taurocholate pancreatitis. Gut 1999;44:253–258.

11 Steinle AU, Weidenbach H, Wagner M, Adler G, Schmid RM: NF-kappaB/Rel activation in cerulein pancreatitis. Gastroenterology 1999;116:420–430.

12 Algul H, Tando Y, Beil M, Weber CK, Von Weyhern C, Schneider G, Adler G, Schmid RM: Different modes of NF-kappaB/Rel activation in pancreatic lobules. Am J Physiol Gastrointest Liver Physiol 2002;283:G270–G281.

13 Tando Y, Algul H, Wagner M, Weidenbach H, Adler G, Schmid RM: Caerulein-induced NF-kappaB/Rel activation requires both Ca^{2+} and protein kinase C as messengers. Am J Physiol 1999;277:G678–G686.

14 Rakonczay Z Jr, Hegyi P, Takacs T, McCarroll J, Saluja AK: The role of NF-kappaB activation in the pathogenesis of acute pancreatitis. Gut 2008;57:259–267.

15 Chen X, Ji B, Han B, Ernst SA, Simeone D, Logsdon CD: Nf-kappaB activation in pancreas induces pancreatic and systemic inflammatory response. Gastroenterology 2002;122:448–457.

16 Aleksic T, Baumann B, Wagner M, Adler G, Wirth T, Weber CK: Cellular immune reaction in the pancreas is induced by constitutively active IkappaB kinase-2. Gut 2007;56:227–236.

17 Baumann B, Wagner M, Aleksic T, von Wichert G, Weber CK, Adler G, Wirth T: Constitutive IKK2 activation in acinar cells is sufficient to induce pancreatitis in vivo. J Clin Invest 2007;117:1502–1513.

18 Algul H, Treiber M, Lesina M, Nakhai H, Saur D, Geisler F, Pfeifer A, Paxian S, Schmid RM: Pancreas-specific RelA/p65 truncation increases susceptibility of acini to inflammation-associated cell death following cerulein pancreatitis. J Clin Invest 2007;117:1490–1501.

19 Vasseur S, Folch-Puy E, Hlouschek V, Garcia S, Fiedler F, Lerch MM, Dagorn JC, Closa D, Iovanna JL: P8 improves pancreatic response to acute pancreatitis by enhancing the expression of the anti-inflammatory protein pancreatitis-associated protein I. J Biol Chem 2004;279:7199–7207.

20 Hezel AF, Kimmelman AC, Stanger BZ, Bardeesy N, Depinho RA: Genetics and biology of pancreatic ductal adenocarcinoma. Genes Dev 2006;20:1218–1249.

21 Naugler WE, Karin M: Nf-kappab and cancer-identifying targets and mechanisms. Curr Opin Genet Dev 2008;18:19–26.

22 Algul H, Adler G, Schmid RM: NF-kappaB/Rel transcriptional pathway: implications in pancreatic cancer. Int J Gastrointest Cancer 2002;31:71–78.

23 Chandler NM, Canete JJ, Callery MP: Increased expression of NF-kappa B subunits in human pancreatic cancer cells. J Surg Res 2004;118:9–14.

24 Wang W, Abbruzzese JL, Evans DB, Larry L, Cleary KR, Chiao PJ: The nuclear factor-kappa B RelA transcription factor is constitutively activated in human pancreatic adenocarcinoma cells. Clin Cancer Res 1999;5:119–127.

25 Liptay S, Weber CK, Ludwig L, Wagner M, Adler G, Schmid RM: Mitogenic and antiapoptotic role of constitutive NF-kappaB/Rel activity in pancreatic cancer. Int J Cancer 2003;105:735–746.

26 Mayo MW, Wang CY, Cogswell PC, Rogers-Graham KS, Lowe SW, Der CJ, Baldwin AS Jr: Requirement of NF-kappaB activation to suppress p53-independent apoptosis induced by oncogenic ras. Science 1997;278:1812–1815.

27 Xiong HQ, Abbruzzese JL, Lin E, Wang L, Zheng L, Xie K: NF-kappaB activity blockade impairs the angiogenic potential of human pancreatic cancer cells. Int J Cancer 2004;108:181–188.

28 Fujioka S, Niu J, Schmidt C, Sclabas GM, Peng B, Uwagawa T, Li Z, Evans DB, Abbruzzese JL, Chiao PJ: Nf-kappab and AP-1 connection: Mechanism of NF-kappaB-dependent regulation of AP-1 activity. Mol Cell Biol 2004;24:7806–7819.

29 Pan X, Arumugam T, Yamamoto T, Levin PA, Ramachandran V, Ji B, Lopez-Berestein G, Vivas-Mejia PE, Sood AK, McConkey DJ, Logsdon CD: Nuclear factor-kappaB p65/RelA silencing induces apoptosis and increases gemcitabine effectiveness in a subset of pancreatic cancer cells. Clin Cancer Res 2008;14:8143–8151.

30 Farrell JJ, Elsaleh H, Garcia M, Lai R, Ammar A, Regine WF, Abrams R, Benson AB, Macdonald J, Cass CE, Dicker AP, Mackey JR: Human equilibrative nucleoside transporter 1 levels predict response to gemcitabine in patients with pancreatic cancer. Gastroenterology 2009;136:187–195.

31 Greten FR, Weber CK, Greten TF, Schneider G, Wagner M, Adler G, Schmid RM: Stat3 and NF-kappaB activation prevents apoptosis in pancreatic carcinogenesis. Gastroenterology 2002;123:2052–2063.

32 Guerra C, Schuhmacher AJ, Canamero M, Grippo PJ, Verdaguer L, Perez-Gallego L, Dubus P, Sandgren EP, Barbacid M: Chronic pancreatitis is essential for induction of pancreatic ductal adenocarcinoma by K-Ras oncogenes in adult mice. Cancer Cell 2007;11:291–302.

33 Hingorani SR, Petricoin EF, Maitra A, Rajapakse V, King C, Jacobetz MA, Ross S, Conrads TP, Veenstra TD, Hitt BA, Kawaguchi Y, Johann D, Liotta LA, Crawford HC, Putt ME, Jacks T, Wright CV, Hruban RH, Lowy AM, Tuveson DA: Preinvasive and invasive ductal pancreatic cancer and its early detection in the mouse. Cancer Cell 2003;4:437–450.

Hana Algül, MD, MPH
II. Medizinische Klinik, Klinikum rechts der Isar,
Technical University of Munich
Ismaninger Strasse 22, DE–81675 Munich (Germany)
Tel. +49 89 4140 6792, Fax +49 89 4140 6795, E-Mail hana.alguel@lrz.tum.de

Mayerle J, Tilg H (eds): Clinical Update on Inflammatory Disorders of the Gastrointestinal Tract.
Front Gastrointest Res. Basel, Karger, 2010, vol 26, pp 176–185

Immunotherapy of Pancreatic Carcinoma: Recent Advances

Angela Märten[a,b] · Markus W. Büchler[a]

[a]Department of Surgery and [b]National Center for Tumor Diseases, University of Heidelberg, Heidelberg, Germany

Abstract

Patients with carcinoma of the exocrine pancreas have an especially poor prognosis. They suffer from a systemic disease that is insensitive to radiotherapy and often to chemotherapy as well. One promising treatment strategy is immunotherapy. This review describes the idea behind chemo-immunotherapy and highlights 3 clinical trials dealing with antibody approaches based on antibody-dependent cellular cytotoxicity, direct cytotoxicity or immunostimulation. Furthermore, 3 clinical trials using immunomodulation and/or vaccination are presented and discussed. Phase II clinical trials investigating combination therapies are promising, although the data have to be discussed carefully until confirmatory phase III data are available. Translational research should always accompany clinical trials to allow further insight into the mode of action and to develop predictive markers.

Patients with carcinoma of the exocrine pancreas have an especially poor prognosis, with a 5-year survival rate of less than 1% and a median survival of 4–6 months. Even after surgical intervention and care at specialized cetres, the 5-year survival rate is at best 25 and 15% with and without adjuvant chemotherapy, respectively [1]. In contrast to other malignancies, pancreatic cancer is often highly resistant to chemotherapy; targeted therapy and the molecular mechanisms that determine treatment resistance are poorly understood. One possible explanation for the unsatisfying results of chemo- and targeted therapy is the strong desmoplastic reaction of pancreatic tumours and the resulting diminished biodistribution of the agents. Immunotherapy might be an option in combination with chemotherapy. Immune cells can recognize tumour cells in a specific way and actively migrate to the tumour, whereas acute inflammatory processes increases the permeability of tumour endothelium and induce higher blood flow transporting active substances. Chemotherapy and immunotherapy act synergistically and not antagonistically [2]. Chemotherapy causes: (1) tumour cell destruction, which leads to the delivery of a broad range of tumour antigens to the immune system [3]; (2) reduction of the immunosuppressive ability of tumours because of tumour debulking (i.e. a reduction of tumour mass) [4], and (3) lymphopenia, which destroys regulatory T cells and allows the infiltration of tumour-reactive cells [3, 4]. Immunotherapy can also boost spontaneous responses to the cross-presentation of endogenous antigens from tumour cells destroyed by chemotherapy, without the requirement for specific peptides, vaccinations or other laborious strategies [5]. The timing of treatment administration is crucial; chemotherapy in parallel to immunotherapy or closely followed by immunotherapy appears to be most effective [6].

As a result of their malignant aberrations, cancer cells differ from normal cells. These changes could serve as potential targets for antibodies, natural killer (NK) cells, or cytotoxic T lymphocytes [7]. T cells appear to be the main effector cells of immune surveillance while the innate immune system, which includes NK cells, macrophages, monocytes, and the complement system, is also involved [8]. Dendritic cells (DCs) are professional antigen-presenting cells (APCs) that function as a bridge, connecting components of the adaptive and innate immune systems. However, the immune system exhibits an anergy or specific pathological tolerance of cancer cells [9]. All cells, including tumour cells, bear MHC molecules on their surface that can display peptides for immune surveillance [10]; however, tumour cells and particularly metastatic cells down-regulate MHC class I molecules, which allows them to escape the immune system [11]. NK cells have long been recognized for their importance in innate immunity, viral defence, and tumour surveillance. In the absence of MHC class I antigen expression – i.e. 'missing self' – on their target cells they induce cytolysis by binding to NKG2DL, a NK cell ligand [12]. Cancer cells can evade immunosuppression by down-regulating intrinsic immunogenicity [13], and stimulation of the immune system has been shown to overcome the unresponsiveness and anergy observed in cancer patients [14].

Immunotherapeutical Approaches

Currently, several approaches are under clinical investigation (table 1). They can be divided into antibody-based strategies and immunomodulatory/vaccination approaches. Antibody-based approaches have to deal with specificity, avidity of the antibody on one hand and the antigen expression on the other. Next, the mode of action should be clarified: (1) antibody-dependent cellular cytotoxicity (ADCC) and/or complement dependent cytotoxicity; (2) antagonistic; (3) agonistic, or (4) blockage of ligands. Here, it is important to point to the right choice of in vivo models for preclinical evaluation. For example, subcutaneous xenograft models in immunocompromised mice are not appropriate for investigation of antibodies acting via ADCC or influencing metastatic spread.

Morab-009 – a chimeric antibody – is currently tested in combination with gemcitabine in a randomized, double-blind phase II trial in Europe and US. MORAb-009

Table 1. Overview on clinical trials in immunotherapy of pancreatic carcinoma

Phase	Condition	Patients n	Intervention	Toxicity	Result	Ref.
III RCT	Adjuvant	110	Chemoradiation plus IFN-α vs. chemotherapy	Haematological toxicity; flu-like symptoms (IFN-α)	Follow-up ongoing. First immunomonitoring data indicate immunomodulatory effects	[26, 30]
II	Adjuvant	80	Chemoradiation plus IFN-α	42% hospitalization due to GI toxicity	Median survival 27.1 months; 2-year overall survival 58%	[29]
II	Adjuvant	43	Chemoradiation plus IFN-α	93% grade 3–4 GI toxicity	Actuarial overall survival for the 1-, 2- and 5-year periods is 95, 64 and 55%	[24]
II	Adjuvant	60	Chemoradiation plus allogeneic GM-CSF tumour vaccine	No dose-limiting toxicities	Follow-up ongoing. Median survival ~26 months. Mesothelin-specific lymphocytes in subjects who demonstrated evidence of clinical responses	[22]
I/II	Advanced	10	MALP-2 intratumorally followed by chemotherapy	20 µg were well tolerated	Actual mean survival 19.6 months; median survival 13.4 months	[23]
I	Advanced	14	Chemoradiation plus allogeneic GM-CSF tumour vaccine	No dose-limiting toxicities	DTH responses to autologous tumour cells in 3 patients accompanied by an increased disease-free survival time (>25 months)	[21]
I	Advanced solid tumours	3 with pancreatic carcinoma	MORAb-009	No dose-limiting toxicities	1 of 3 subjects showed stable disease by CT and drop in CA 19-9 whilst progressive under gemcitabine	[15]
I	Advanced solid tumours	NA	^{90}Y-hPAM4	No dose-limiting toxicities	Several patients had objective evidence of tumour shrinkage	[17] (personal communication)
I	Advanced solid tumours	29	CP-870,893	Grade 1–2 cytokine release syndrome	Marked, dose-related up-regulation of co-stimulatory and MHC molecules on B cells	[20]

DTH = Delayed-type hypersensitivity; MALP-2 = macrophage-activating lipopeptide-2; PC = pancreatic carcinoma; GM-CSF = granulocyte-macrophage-colony stimulating factor.

is a monoclonal antibody that targets mesothelin, a cell surface adhesive protein over-expressed in pancreatic carcinoma with minimal expression in normal tissue. It was identified as one of the most prominent, differentially expressed cell surface antigens on pancreatic adenocarcinoma using SAGE of primary tumours versus normal pancreatic tissue. MORAb-009 is effective in mediating cell killing by ADCC and inhibiting cell adhesion. MORAb-009 has single agent anti-tumour activity which could be enhanced by the addition of chemotherapy. The toxicity profile from the phase I trial showed no dose limiting toxicity and a good safety profile. One of 3 subjects with pancreatic cancer enrolled in the phase I trial progressed on gemcitabine but showed stable disease by CT and a drop in CA 19-9 after MORAb-009 infusion [15].

^{90}Y-hPAM4 – a CDR-grafted, humanized monoclonal IgG1 antibody specific for the MUC1 antigen – is currently tested in combination with gemcitabine in a phase Ib trial in Europe and the USA. PAM4 is an antibody produced against the MUC1 mucin glycoprotein isolated from xenografted human pancreatic cancer. Most pancreatic cancers are mucin-producing adenocarcinomas and initial characterization studies showed PAM4 reacted with approximately 85% of pancreatic cancers [16]. Initial immunoscintigraphic studies using radiolabelled antibody demonstrated specific targeting in patients with pancreatic cancer subsequently confirmed at surgery. In tumour-bearing animals, Gold et al. [17] reported that PAM4 radiolabelled with I-131 or Y-90 has growth-inhibitory effects against pancreatic cancer xenografts, that radiolabelled PAM4 is more potent than gemcitabine when compfared at maximal tolerated dose, and that combining radiolabelled PAM4 with gemcitabine further increased anti-tumour activity and survival results. These studies also duplicated repeated cycles of gemcitabine as given clinically, adding the ^{90}Y-PAM4 to 3 of the cycles in a repeated dosing, which proved to be the most effective regimen. The first clinical trial of ^{90}Y-hPAM4 was a single-agent dose-escalation trial in patients with unresectable locally advanced or metastatic disease most of whom had failed 5-FU, gemcitabine or other standard therapy. Several patients had objective evidence of tumour shrinkage.

Vonderheide et al. evaluated the dynamics of the immune reaction to pancreatic cancer from inception to invasion [18]. They showed that in a genetically defined mouse model of pancreatic ductal adenocarcinoma, suppressive cells of the host immune system appear early during pancreatic tumorigenesis preceding and overcoming anti-tumour cellular immunity, and are likely to contribute to disease progression [18]. Therefore, strategies which stimulate the immune system (i.e. shift the balance from suppression towards activation) are of particular interest. Furthermore, few tumour-associated antigens have been described; therefore, strategies aimed at 'self tumour-antigen delivery' in vivo are especially interesting. A strategy in which a CD40L encoding plasmid was delivered intraperitoneally was shown to retard tumour-growth in an orthotopic, syngeneic mouse model [5]. CD40 is expressed on APCs, including DCs, B cells, activated macrophages, and follicular DCs. CD40-CD40L interactions play a pivotal role in the activation of professional APCs. CD40L enhances antigen presentation of CD40-expressing APCs and activation of effector T cells and NK cells. This interaction leads to an up-regulation of adhesion molecules on endothelial cells and stimulation of NK cells, CD4+ T cells and DCs [19]. It is likely that CD40L-expressing cells migrate from the plasmid injection site to a draining lymph node into which the pancreas also drains. In the lymph node, the CD40L-expressing cells from the plasmid injection site could directly stimulate CD40 on the surface of tumour associated antigen-bearing DCs from the pancreas. Another possible mechanism might involve the release of inflammatory factors from the plasmid injection site, which could activate DCs from the pancreas. Furthermore, CD40 is also expressed by 30–70% of solid tumours, and engagement of CD40 on tumour cells results in apoptosis. CP-870,893 – a fully human and selective CD40 agonist monoclonal antibody – is currently being tested in combination with gemcitabine in a phase I–II trial in Europe and the USA. CP-870,893 activates human APC in vitro and inhibits growth of human tumours in immune-deficient and immune-reconstituted SCID-beige mice. In a phase I dose-escalation study, CP-870,893 was administered i.v. once in patients with advanced solid tumours (including pancreatic cancer). The most common adverse event was grade 1–2 cytokine release syndrome. Four out of 11 patients with melanoma had objective partial responses. CP-870,893 infusion resulted in depletion of peripheral CD19+ B cells (>80% depletion at the highest dose levels). Among B cells remaining in blood, a marked, dose-related up-regulation of co-stimulatory and MHC molecules after treatment was described [20].

Promising immunomodulatory/vaccination protocols are currently under investigation in clinical trials. Jaffee et al. [14] demonstrated in preclinical studies with murine tumour models that tumour cell vaccines engineered to secrete GM-CSF in a paracrine manner elicited systemic immune responses capable of eliminating small amounts of established pancreatic tumour. In a phase I clinical safety trial, pancreatic carcinoma patients (n = 14) were treated with a gene-modified granulocyte-macrophage colony stimulating factor (GM-CSF) secreting allogeneic pancreatic cancer cell vaccine in combination with chemoradiotherapy (5-FU plus radiotherapy) [21]. Safety was demonstrated and clinical benefit, particularly in patients exhibiting an immune response, was observed [21]. In 2005, a phase II clinical trial (60 patients with resected pancreatic adenocarcinoma) to asses the above described treatment protocol had completed enrolment. The first data reported from this trial are encouraging, with a 2-year survival of 70% and a median survival of approximately 26 months [22]. In comparison, a median survival of 20–22 months was observed when the standard adjuvant treatment protocols with the chemotherapeutic agents 5-FU or gemcitabine were administered. In addition, immunomonitoring identified mesothelin-specific T lymphocytes in patients who demonstrated immune and clinical responses. The trial completed enrolment in February 2008.

Some clinical trials tested local (intratumoral) administration of the investigational agent [23]. Local administration of immuno-active substances has been shown to induce local inflammation and direct leucocytes to the tumour [23]. At the tumour site, APCs might engulf apoptotic or necrotic tumour cells and prime T lymphocytes, which could then infiltrate the tumour and induce tumour death. The TLR2/TLR6 agonist macrophage-activating lipopeptide-2 (MALP-2) in combination with gemcitabine was investigated in a phase I/II clinical trial in patients with incompletely resectable pancreas carcinomas (n = 10) [23]. Patients were injected intratumorally during surgery with MALP-2 (20–30 μg) followed by post-operative chemotherapy. The 20-μg dose was well tolerated and signs of local MALP-2 effects were observed; lymphocytes and monocytes were detected in wound secretions and tumour-induced inhibition of NK cells was abolished. The actual mean survival is 19.6 months and the median survival 13.4 months [23], one patient is still alive (>50 months after laparotomy).

In 1995, a phase II clinical trial in patients with high-risk resected pancreatic adenocarcinoma (n = 43; 84% node positive and 19% margin positive) was initiated by the Virginia Mason Clinic (Seattle, USA) to investigate pancreaticoduodenectomy and adjuvant therapy with 5-FU, cisplatin, IFN-α and radiation therapy. After a median follow-up period of 32 months, the 2-year survival rate was 64% and the 5-year survival rate was 55% [24]. In addition, the overall recurrence rate was 12%, of which 80% occurred within 2 years following surgery. Data from approximately 100 patients with high-risk resected pancreatic adenocarcinoma showed a 5-year survival rate of 50% [Picozzi, pers. commun.]. In comparison, conventional treatment regimens demonstrate a 5-year survival rate of approximately 21%. From August 2004 to December 2007, a phase III randomized clinical trial to compare the Virginia Mason Clinic scheme with standard chemotherapy recruited 100 patients in Germany and Italy [25]. The adjuvant chemo-radio-immunotherapy of pancreatic carcinoma (CapRI) trial is an open-label, controlled, prospective, randomized multicentre trial to evaluate the post-operative overall survival of patients with pancreatic adenocarcinoma that received chemoradiotherapy, including IFN-α2b administration, compared with adjuvant chemotherapy [25]. IFN-α was included in this treatment strategy because it has been shown to increase the effectiveness of chemoradiotherapy [26] (fig. 1). In vitro and in vivo data suggest that immunostimulation is the mechanism of improved efficacy of this treatment regimen [27]; therefore, immunomonitoring in this trial focused on NK cell activity, immunophenotyping of peripheral blood, an analysis of tumour-specific cytokine release, tumour lymphocyte infiltration, and an analysis of various serum markers. Strong immune reactions were elicited by IFN-α; a robust increase in monocytes, peripheral DCs, CD40 expression, central memory T cells, and NK cell-mediated cytotoxicity was observed 1 day following the first IFN-α injection. After 4 days, the serum levels of TNF-α and IL-12 peaked. After 6 weeks a second peak in peripheral DC, an increase of effector memory T cells in parallel with the multimodality treatment phase and, most interestingly, antigen-specific T cells were observed. These observations could be explained by a series of events that begin with the immune system's attack on the tumour, initiated and maintained by the injection of IFN-α and leading to a switch from innate to specific immune system responses

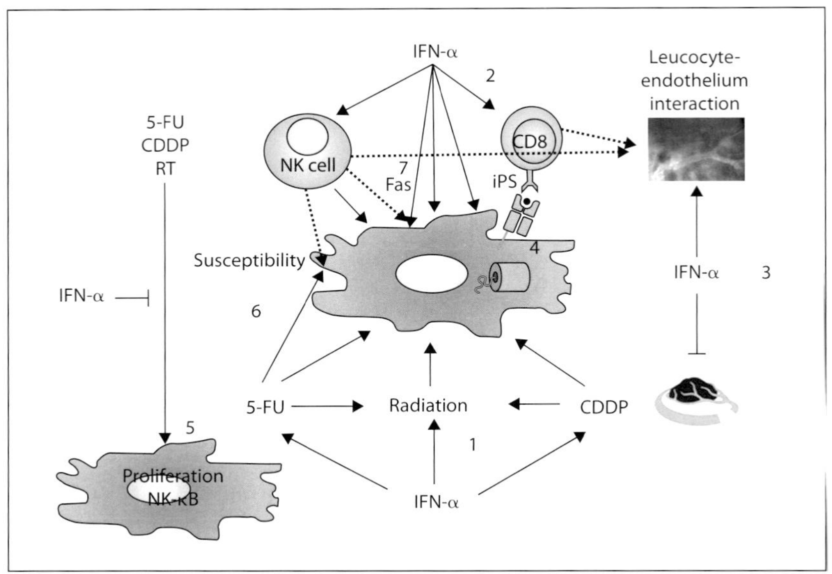

Fig. 1. Mechanism of IFN-α alone and in combination with chemo- and radiotherapy. 5-FU, cisplatin and IFN-α are radiosensitizers (1). IFN-α is also a chemosensitizer (1). IFN-α enhances NK cell mediated cytotoxicity and, to a lesser extent, CD8 cells (2). IFN-α has anti-angiogenic action and enhances leucocyte-endothelium interactions (3). Treatment with IFN-α induces the switch to the immunoproteasome resulting in an increased immunogenicity (4). Cisplatin induces rapid cell re-growth; this could be prevented by IFN-α (5). Chemo- as well as radiotherapy induces NF-κB, known to promote growth of pancreatic carcinoma. IFN-α inhibits NF-κB up-regulation or even down-regulates NF-κB (5). 5-FU makes pancreatic tumour cells more susceptible to IFN-α triggered NK cell attacks (6). IFN-α induces up-regulation of Fas, thus resulting in an increase in NK cell cytotoxicity (7). CDDP = Cisplatin; RT = radiotherapy.

[28]. Interestingly, preliminary data suggest a correlation between NK cell activity, appearance of central memory T cells and clinical outcome [unpubl. data from the author]. First data for clinical efficacy are expected at the end of 2009. Toxicities observed are moderate and mainly of haematological origin, which is in contrast to the daunting safety data reported from the Virginia Mason Clinic and the ACOSOG group [29]. It seems that differences in supportive therapy and definition of radiation fields might be responsible for the differences. Nevertheless, de-escalated regimens for reducing toxicity are now investigated in the 3-armed, randomized, multicentre CapRI-2 trial. A de-escalation of the CapRI-scheme will be tested in 2 different modifications. Patients in study arm A will be treated as outpatients with the complete CapRI-scheme consisting of cisplatin, IFN-α2b and external beam radiation and 3 cycles of 5-FU continuous infusion. In study arm B the first de-escalation will be realized by omitting cisplatin. Next, patients in study arm C will additionally not receive external beam radiation. A total of 135 patients with pathologically confirmed R0 or R1 resected pancreatic adenocarcinoma are planned to be enrolled.

Summary

Ductal adenocarcinoma of the pancreas is one of the leading causes of cancer death in the UK, Europe and USA, with incidence closely paralleling mortality. However, it has been recognized that specialist surgery can improve outcome and there is high level evidence that adjuvant chemotherapy offers a significant advantage in terms of 5-year survival. One further strategy is immunotherapy, for which early phase clinical trials have shown promising data. Phase II clinical trials investigating combination therapies are highly promising and a randomized phase III clinical trial is investigating a combination of chemoradiotherapy and IFN-α, the only immunotherapeutic strategy currently under investigation [25]. One major focus of immunotherapeutic interventions is to monitor the immune response. Various techniques for immunomonitoring have been established and in addition to the detection of antigen-specific T cells with a defined epitope, samples can be analysed for cells derived from the innate immune system and for humoral responses. Immunoassays or tests that measure a pre-designated immunological parameter are useful, but require adherence to standard operating procedures, standardization and validation to be deemed reliable. Assay validation requires an evaluation of assay accuracy, precision, limits of detection, limits of quantitation, specificity, linearity and range robustness, and system suitability. Toxicities observed in the immunotherapy clinical trials were mostly mild to moderate and included 'classical' flu-like symptoms and fever. The data suggest that immunotherapy is expected to have a major impact on this multimodality treatment. However, it has to be emphasized that immunotherapy should be combined with chemotherapy or chemoradiotherapy. Data from the CapRI trial is expected to significantly advance clinical and scientific knowledge on the use of adjuvant treatment for the treatment of pancreatic carcinoma and may confirm and elucidate the remarkable data generated by the Virginia Mason study group. Immunotherapy for the treatment of pancreatic carcinomas appears promising with respect to toxicity and efficacy. However, high-quality randomized controlled trials combined with immunomonitoring are required to prove the role of immunotherapy.

References

1 Raraty MG, Magee CJ, Ghaneh P, Neoptolemos JP: New techniques and agents in the adjuvant therapy of pancreatic cancer. Acta Oncol 2002;41:582–595.

2 Lake R, Robinson B: Immunotherapy and chemotherapy: a practical partnership. Nat Rev Cancer 2005;5:397–405.

3 Nowak AK, Lake RA, Marzo AL, Scott B, Heath WR, Collins EJ, Frelinger JA, Robinson BW: Induction of tumor cell apoptosis in vivo increases tumor antigen cross-presentation, cross-priming rather than cross-tolerizing host tumor-specific CD8 T cells. J Immunol 2003;170:4905–4913.

4 Nowak AK, Robinson BW, Lake RA: Gemcitabine exerts a selective effect on the humoral immune response: implications for combination chemo-immunotherapy. Cancer Res 2002;62:2353–2358.

5 Serba S, Schmidt J, Wentzensen N, Ryschich E, Märten A: Transfection with CD40L induces tumour suppression by dendritic cell activation in an orthotopic mouse model of pancreatic adenocarcinoma. Gut 2008;57:344–351.
6 Lake RA, Robinson BW: Immunotherapy and chemotherapy: a practical partnership. Nat Rev Cancer 2005;5:397–405.
7 Pure E, Allison JP, Schreiber RD: Breaking down the barriers to cancer immunotherapy. Nat Immunol 2005;6:1207–1210.
8 Janeway C, Travers P, Walport M, Shlomchik M: Immunobiology, ed 5. New York, Garland Publishing, 2001.
9 Gajewski TF, Meng Y, Harlin H: Immune suppression in the tumor microenvironment. J Immunother 2006;29:233–240.
10 Zinkernagel RM: On cross-priming of MHC class I-specific CTL: rule or exception? Eur J Immunol 2002;32:2385–2392.
11 Khong HT, Restifo NP: Natural selection of tumor variants in the generation of 'tumor escape' phenotypes. Nat Immunol 2002;3:999–1005.
12 Takeda K, Okumura K: CAM and NK Cells. Evid Based Complement Alternat Med 2004;1:17–27.
13 Bissell MJ, Radisky D: Putting tumours in context. Nature Rev Cancer 2001;1:46–54.
14 Jaffee EM, Hruban RH, Biedrzycki B, Laheru D, Schepers K, Sauter PR, Goemann M, Coleman J, Grochow L, Donehower RC, Lillemoe KD, O'Reilly S, Abrams RA, Pardoll DM, Cameron JL, Yeo CJ: Novel allogeneic granulocyte-macrophage colony-stimulating factor-secreting tumor vaccine for pancreatic cancer: a phase I trial of safety and immune activation. J Clin Oncol 2001;19:145–156.
15 Armstrong D, Laheru D, Ma W, Cohen S, Phillips M, Brahmer J, Weil S, Hassan R: A phase 1 study of MORAb-009, a monoclonal antibody against mesothelin in pancreatic cancer, mesothelioma and ovarian adenocarcinoma. J Clin Oncol ASCO Ann Meet Proc 2007;25:14041.
16 Gold DV, Lew K, Maliniak R, Hernandez M, Cardillo T: Characterization of monoclonal antibody PAM4 reactive with a pancreatic cancer mucin. Int J Cancer 1994;57:204–210.
17 Gold DV, Schutsky K, Modrak D, Cardillo TM: Low-dose radioimmunotherapy [(90)Y-PAM4] combined with gemcitabine for the treatment of experimental pancreatic cancer. Clin Cancer Res 2003;9:3929S–3937S.
18 Clark CE, Hingorani SR, Mick R, Combs C, Tuveson DA, Vonderheide RH: Dynamics of the immune reaction to pancreatic cancer from inception to invasion. Cancer Res 2007;67:9518–9527.
19 Costello RT, Gastaut JA, Olive D: What is the real role of CD40 in cancer immunotherapy? Immunol Today 1999;20:488–493.
20 Vonderheide R, Flaherty K, Khalil M, Stumacher M, Bajor D, Gallagher M, Sullivan P, Mahaney J, O'Dwyer P, Huhn R, Antonia S: Clinical activity and immune modulation in cancer patients treated with CP-870,893, a novel CD40 agonist monoclonal antibody. J Clin Oncol ASCO Ann Meet Proc 2006;24:2507.
21 Jaffee EM, Abrams R, Cameron J, Donehower R, Duerr M, Gossett J, Greten TF, Grochow L, Hruban R, Kern S, Lillemoe KD, O'Reilly S, Pardoll D, Pitt HA, Sauter P, Weber C, Yeo C: A phase I clinical trial of lethally irradiated allogeneic pancreatic tumor cells transfected with the GM-CSF gene for the treatment of pancreatic adenocarcinoma. Hum Gene Ther 1998;9:1951–1971.
22 Laheru D, Yeo C, Biedrzycki B, Solt S, Lutz E, Onners B, Tartakovsky I, Herman J, Hruban R, Piantadosi S, Jaffee EM: A safety and efficacy trial of lethally irradiated allogeneic pancreatic tumor cells transfected with the GM-CSF gene in combination with adjuvant chemoradiotherapy for the treatment of adenocarcinoma of the pancreas. J Clin Oncol ASCO Ann Meet Proc 2007;25:3010.
23 Schmidt J, Welsch T, Jäger D, Mühlradt PF, Büchler MW, Märten A: Intratumoral injection of the toll-like receptor-2/6 agonist 'macrophage-activating lipopeptide-2' in patients with pancreatic carcinoma: a phase I/II trial. Br J Cancer 2007;97:598–604.
24 Picozzi VJ, Kozarek RA, Traverso LW: Interferon-based adjuvant chemoradiation therapy after pancreaticoduodenectomy for pancreatic adenocarcinoma. Am J Surg 2003;185:476–480.
25 Knaebel HP, Marten A, Schmidt J, Hoffmann K, Seiler C, Lindel K, Schmitz-Winnenthal H, Fritz S, Herrmann T, Goldschmidt H, Mansmann U, Debus J, Diehl V, Buchler MW: Phase III trial of postoperative cisplatin, interferon alpha-2b, and 5-FU combined with external radiation treatment versus 5-FU alone for patients with resected pancreatic adenocarcinoma: CapRI: study protocol [ISRCTN62866759]. BMC Cancer 2005;5:37.
26 Schmidt J, Patrut EM, Ma J, Jäger D, Knaebel HP, Büchler MW, Märten A: Immunomodulatory impact of interferon-alpha in combination with chemoradiation of pancreatic adenocarcinoma (CapRI). Cancer Immunol Immunother 2006;55:1396–1405.
27 Zhu Y, Tibensky I, Schmidt J, Ryschich E, Marten A: Interferon-alpha enhances antitumor effect of chemotherapy in an orthotopic mouse model for pancreatic adenocarcinoma. J Immunother 2008;31:599–606.

28 Schmidt J, Jäger D, Hoffmann K, Büchler MW, Märten A: Impact of interferon-alpha in combined chemoradioimmunotherapy for pancreatic adenocarcinoma (CapRI): first data from the immunomonitoring. J Immunother 2007;30:108–115.

29 Picozzi J, Abrams R, Traverso L, O'Reilly E, Greeno E, Martin R, Wilfong L, Decker P, Pisters P, Posner M: ACOSOG Z05031: Report on a multicenter, phase II trial for adjuvant therapy of resected pancreatic cancer using cisplatin, 5- FU, and alpha-interferon. J Clin Oncol 2008;26(suppl):abstr 4505.

30 Knaebel HP, Märten A, Schmidt J, Hoffmann K, Seiler C, Lindel K, Schmitz-Winnenthal H, Fritz S, Herrmann T, Goldschmidt H, Krempien R, Mansmann U, Debus J, Diehl V, Büchler MW: Phase III trial of postoperative cisplatin, interferon alpha-2b, and 5-FU combined with external radiation treatment versus 5-FU alone for patients with resected pancreatic adenocarcinoma. CapRI: study protocol [ISRCTN62866759]. BMC Cancer 2005;5:37.

Angela Märten, PhD
Department of Surgery
Im Neuenheimer Feld 350
DE–69120 Heidelberg (Germany)
Tel. +49 622 156 39890, Fax +49 622 156 8240, E-Mail angela.maerten@med.uni-heidelberg.de

Mayerle J, Tilg H (eds): Clinical Update on Inflammatory Disorders of the Gastrointestinal Tract.
Front Gastrointest Res. Basel, Karger, 2010, vol 26, pp 186–198

Helicobacter pylori Infection: To Eradicate or Not to Eradicate

Kerstin Schütte · Arne Kandulski · Michael Selgrad · Peter Malfertheiner

Department of Gastroenterology, Hepatology and Infectious Diseases, Otto von Guericke University Magdeburg, Magdeburg, Germany

Abstract

The discovery of *Helicobacter pylori* signalled a turning point in our understanding of gastroduodenal pathology. While the infection with *H. pylori* leads to chronic gastritis in all affected individuals, the disease does not progress to any complication in 80% of patients. However, in a subset of patients *H. pylori* causes peptic ulcer disease, mucosa-associated lymphoid tissue lymphoma and gastric adenocarcinoma. While there is a clear indication to eradicate the infection with *H. pylori* in every patient with an associated complication, the question of eradication therapy for preventive purposes is debated. Nevertheless, *H. pylori* eradication before the development of gastric atrophy or intestinal metaplasia has the best potential to prevent gastric cancer. The challenge for the future will not only be to cure peptic ulcer disease and to prevent gastric cancer by eradication of *H. pylori* but to prevent even the infection itself.

The discovery of *Helicobacter pylori* over 25 years ago [1] signalled a turning point in our understanding of gastroduodenal pathology. Chronic active gastritis is initiated in all subjects infected with *H. pylori* and, in its various phenotypic expressions, it is the basis for the development of several diverse complications. Still, it needs to be mentioned that around 80% of all infected patients will not progress to any complication. This review gives an overview on the role of *H. pylori* in gastrointestinal diseases, including peptic ulcer disease and ulcers associated with non-steroidal anti-inflammatory drugs, gastro-oesophageal reflux disease (GERD), functional dyspepsia, mucosa-associated lymphoid tissue (MALT) lymphoma and gastric adenocarcinoma as well as in some extragastrointestinal diseases such as iron deficiency anaemia and idiopathic thrombocytopenic purpura. While *H. pylori* must be eradicated in patients with related complications (table 1), the question whether to eradicate for the purpose of prevention is still a matter of debate.

Table 1. Indications for *H. pylori* eradication according to the Maastricht III Consensus Report [4]

Recommendation	Level of evidence	Grade of recommendation
Eradication is an appropriate option for patients infected with *H. pylori* and investigated non-ulcer dyspepsia	1a	A
Test and treat is an appropriate option for patients with uninvestigated dyspepsia	1a	A
Test and treat or empirical acid suppression are appropriate options in populations with a low *H. pylori* prevalence	2a	B
H. pylori does not cause GERD	1b	A
Eradication does not affect the outcome of PPI treatment in patients with GERD in Western populations	1b	A
Routine *H. pylori* testing is not recommended in GERD	1b	A
Testing should be considered for patients receiving long-term maintenance treatment with PPIs	2b	B
H. pylori eradication is inferior to PPI maintenance therapy in patients receiving long-term NSAIDs and who have peptic ulcer and/or ulcer bleeding in preventing ulcer recurrence and/or bleeding	1b	A
Eradication is of value in chronic NSAID users but is insufficient to prevent NSAID-related ulcer disease completely	1b	A
Eradication may prevent peptic ulcer and/or bleeding in naïve users of NSAIDs	1b	A

Peptic Ulcer Disease

The lifetime risk for developing peptic ulcer disease in *H. pylori* infected patients is approximately 15% [2]. However, the clinical outcome of *H. pylori* infection varies depending on host, environmental and bacterial virulence factors. While antral-predominant gastritis shows a strong correlation to the development of duodenal ulcers the risk for developing gastric ulcers, gastric atrophy, intestinal metaplasia and finally adenocarcinoma is higher in patients with corpus-predominant gastritis [3]. Eradication therapy is imperative in *H. pylori* infected patients who have a current or past medical history of peptic ulcer disease [4].

Furthermore, the annual relapse rate for gastric or duodenal ulcer can dramatically be reduced from more than 50% to 0–10% by eradication of *H. pylori* [5].

Besides *H. pylori* infection, NSAIDs significantly and independently increase the risk of peptic ulcer bleeding. Ninety percent of ulcers can be attributed to one or both of these factors [6]. Although the interaction of *H. pylori* and NSAIDs concerning gastrointestinal epithelial damage and the pathogenesis of ulceration is complex and not currently understood in detail, a significantly increased risk of peptic ulcer bleeding in *H. pylori* infected patients compared to *H. pylori* negative subjects on NSAID treatment has been shown by a number of clinical studies. A well-performed systematic review reports an OR of 6.1 (95% CI 3.9–9.6) for NSAID-treated patients infected with the bacterium in comparison to an OR of 4.8 (95% CI 3.8–6.2) for those who are not infected [7].

However, controversy still surrounds the indication for eradication therapy in patients under long-term treatment with NSAIDs due to conflicting results from clinical trials.

Chan et al. [8] performed a prospective randomized controlled trial in 102 patients who were NSAID naïve, *H. pylori* positive, had a past medical history of dyspepsia or ulcer disease and required long-term NSAID treatment. Patients were either treated with omeprazol in combination with amoxicillin and clarithromycin as eradication therapy for 1 week or with omeprazol and placebo antibiotics for 1 week. The 6-month probability of ulcers was significantly lower in the eradication group [12.1% (95% CI 3.1–21.1) vs. 34.4% (95% CI 21.1–47.7), p = 0.0085] and the same held true for the risk of complicated ulcers [4.2% (95% CI 1.3–9.7) vs. 27.1% (95% CI 14.7–39.5%), p = 0.0026]. These results led to the recommendation to screen for and treat *H. pylori* infection in patients before starting long-term NSAID treatment [8]. On the other hand, a prospective randomized controlled trial on 347 patients already on long-term NSAID treatment and positive on *H. pylori* serologic testing that were either treated with *H. pylori* eradication or placebo failed to show a significant difference in the development of gastroduodenal erosions or dyspepsia [9]. This was in concordance with a previous randomized double-blind placebo-controlled trial in *H. pylori* positive patients requiring NSAID therapy without past or present peptic ulcer disease. In that study, patients were assigned to 1 of 4 arms, either receiving eradication therapy followed by placebo, eradication therapy followed by proton pump inhibitor (PPI) therapy, PPI, or placebo for 4 weeks each. All 3 active therapies reduced the occurrence of NSAID-associated peptic ulcer and dyspeptic symptoms requiring therapy after a follow-up of 5 weeks, but additional eradication therapy was not more effective than PPI treatment alone in the primary prevention of ulcers and dyspepsia during short-term treatment in this cohort of *H. pylori* infected patients at low risk [10]. Among patients with *H. pylori* infection and a history of upper gastrointestinal bleeding who are taking low-dose aspirin, the eradication of *H. pylori* is equivalent to treatment with omeprazole in preventing recurrent bleeding. However, in patients who are taking other NSAIDs, PPI therapy has been shown to be superior to eradication therapy [11]. On the other hand, a recent study showed that *H. pylori* eradication in patients on long-term NSAID therapy leads to healing of gastritis despite ongoing NSAID therapy [12].

H. pylori eradication is of value in chronic NSAID users but is insufficient to completely prevent NSAID-related ulcer disease. In naïve NSAID users, *H. pylori* eradication may prevent peptic ulcer and bleeding. In patients who are receiving long-term aspirin and who bleed, 'test and treat' is the recommended strategy [4].

Gastro-oesophageal Reflux Disease

While some epidemiological studies reported a negative association between GERD and *H. pylori* infection, implying a protective effect of the infection against GERD, other studies did not reveal a negative effect of *H. pylori* eradication on the occurrence or intensity of reflux symptoms [13, 14]. There is a lower prevalence of *H. pylori* in patients with GERD than in patients without reflux disease but the nature of this relationship is uncertain [15]. The possible association of *H. pylori* with GERD is most likely related to the modification of the acidity of gastric juice as a consequence of variable patterns of gastritis. The competence of the gastro-oesophageal junction or oesophageal acid clearance mechanisms are not altered by the infection [16]. While there are studies showing that *H. pylori* eradication increases oesophageal acid exposure [17] and the incidence of reflux oesophagitis in patients with duodenal ulcers 3 years after *H. pylori* eradication in comparison to those with persistent infection (25.8 vs. 12.9%) [18], this was contradicted by the results of other studies even showing a decreased incidence of heartburn in patients with duodenal ulcer after successful eradication therapy [19]. Meanwhile, several studies showed neither a causation of GERD by *H. pylori* eradication nor an exacerbation of pre-existing symptoms, independently of PPI treatment [20, 21].

In *H. pylori* positive patients, profound acid suppression induces a corpus-predominant pangastritis associated with accelerated corpus gland loss and development of atrophic gastritis [22]. Both conditions are associated with an increased risk for gastric cancer. Patients with GERD often need long-term PPI therapy. Eradication therapy leads to healing of gastritis within 12–24 months [23] accompanied by a certain regression of atrophic gastritis [24]. This is the rationale for the recommendation to consider a test and treat strategy in GERD patients receiving long-term maintenance therapy with PPI [4].

Functional Dyspepsia

Dyspepsia defines a wide spectrum of gastrointestinal disorders that can be either of organic cause or of functional nature. It affects, at least sporadically, up to 25% of the population [25]. The Rome III criteria define functional dyspepsia as persistent or recurrent pain or discomfort centred in the upper abdomen with no evidence of organic causes to explain the symptoms (absence of abnormalities in both upper GI endoscopy and abdominal ultrasound). In the clinical management of patients with

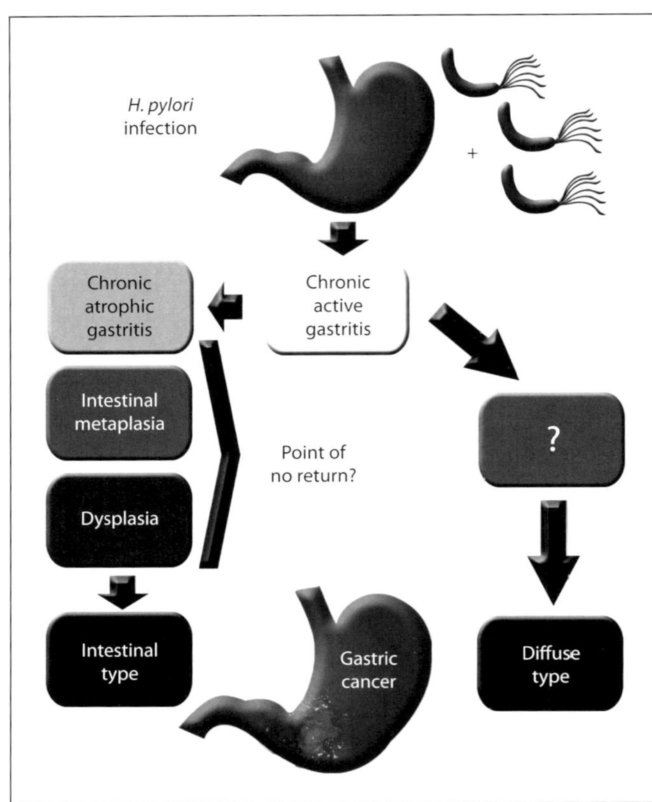

Fig. 1. Multistep process of human gastric carcinogenesis: the Correa sequence [36].

dyspeptic symptoms, those over the age of 45 years or those who present with the first onset of alarm symptoms (odynophagia, anaemia or evidence of acute/chronic bleeding, dysphagia, previous history of peptic ulcer disease, unintentional weight loss or recurrent or persistent vomiting) deserve prompt endoscopy for further investigation [26]. In the remaining patients, *H. pylori* test and treat is considered to be the first-line strategy in areas with high *H. pylori* prevalence (>20%), while in areas with low *H. pylori* prevalence (<20%) an empiric treatment with PPIs is an equivalent option [4] (fig. 2). Test and treat has been validated as beneficial and cost-effective by the randomized placebo-controlled CADET-Hp trial [27].

The urea breath test and the stool antigen tests, both characterized by sensitivity and specificity rates of up to 93–95%, are validated non-invasive diagnostic tools that reliably separate *H. pylori*-positive from *H. pylori*-negative dyspeptic patients in the test and treat strategy [28, 29] (table 2). When comparing the test-and-treat strategy to prompt endoscopy, no differences in symptomatic outcomes or quality of life of patients were detectable even after a follow-up period of 12 months [30]. The test-and-treat strategy does have the advantage of being more cost effective [31].

Table 2. Methods for the detection of *H. pylori* infection, according to [29]

	Sensitivity, %	Specificity, %
Invasive methods		
Culture	70–90	100
Histology	80–98	90–98
Rapid urease test	90–95	90–96
PCR	90–95	90–95
Non-invasive methods		
Urea breath test	85–95	85–95
Monoclonal stool antigen test	85–95	85–95
Serological test for IgG antibodies	70–90	70–90

According to large population studies, *H. pylori* is more frequently detected in dyspeptic patients than in controls [32] but the pathophysiologic role of *H. pylori* in functional dyspepsia is not yet clear. However, *H. pylori* eradication leads to symptom improvement in non-ulcer dyspepsia in addition to reducing the risk of developing peptic ulcers, atrophic gastritis and gastric cancer. In a study by McCarthy et al. [33] on patients with non-ulcer dyspepsia, the investigators demonstrated that successfully treated patients were almost asymptomatic compared to patients with persistent infection or re-infection after a follow-up period of 1 year. Several randomized placebo-controlled trials evaluated the efficacy of *H. pylori* eradication therapy in comparison to antisecretory therapy in patients with functional dyspepsia. A Cochrane database systematic review with a well-performed meta-analysis confirmed the superiority of *H. pylori* eradication in non-ulcer dyspepsia in comparison to either H2 receptor antagonists or sucralfate, with a number needed to treat of 14 patients [34].

H. pylori Therapy for the Prevention of Gastric Cancer

Gastric cancer is a multifactorial disease, but the infection with *H. pylori* is known to be the leading actor in this event. By 1994, the International Agency for Research on Cancer of the World Health Organization had already classified *H. pylori* as a class I human carcinogen. At that time the evidence was based mainly on epidemiological studies showing that the infection with *H. pylori* increases the risk of gastric cancer by a factor of 2–3. More recently, by overcoming methodological flaws, about 70% of non-cardiac gastric adenocarcinomas can be attributed to the infection, with the risk increasing to 20 times for those carrying the *H. pylori* infection [35]. The progression of chronic gastritis, initiated by *H. pylori* infection and perpetuated by environmental and host factors, to atrophic gastritis, intestinal metaplasia and, in some patients, to

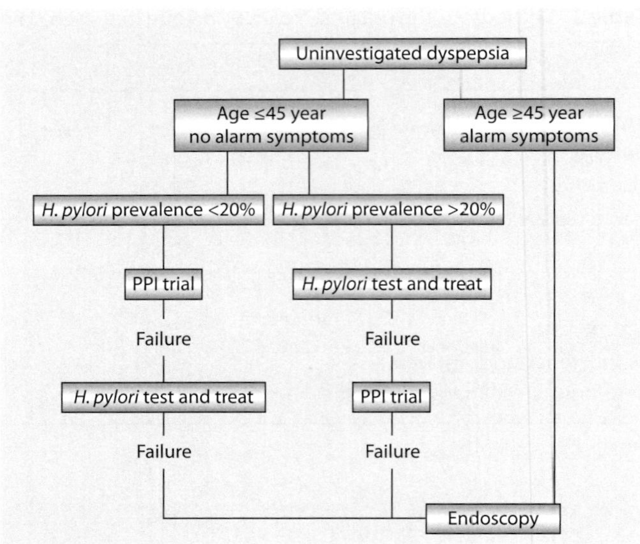

Fig. 2. Algorithm for the management of uninvestigated dyspepsia according to the Maastricht III Consensus Report [4].

dysplasia and finally gastric cancer is known as the Correa cascade [36] (fig. 1). It has been intensely studied in animal models. In Mongolian gerbils the induction of well-differentiated adenocarcinomas by infection with *H. pylori* has been observed and has provided evidence that *H. pylori* is a carcinogen [37, 38]. In a large prospective endoscopic follow-up study on 1,526 patients with duodenal or gastric ulcers, gastric hyperplasia or non-ulcer dyspepsia, of whom 1,246 were infected with *H. pylori*, Uemura et al. [39] showed a development of gastric cancer only in infected patients. A significantly increased risk for gastric cancer was shown in patients with severe gastric atrophy, corpus-predominant gastritis or intestinal metaplasia [39]. A meta-analysis of 19 studies including 2,491 patients and 3,959 controls revealed an OR of 1.92 (95% CI 1.32–2.78) for gastric cancer in *H. pylori* infected patients [40]. Bacterial virulence factors and host genetic factors in addition to certain environmental conditions are relevant in increasing the susceptibility to cancer. In a meta-analysis by Huang et al. [41] on 16 case-control studies, it was shown that an infection with cagA (cytotoxin-associated antigen A)-positive strains of *H. pylori* increases the risk for gastric cancer over the risk associated with *H. pylori* infection alone. *H. pylori* and cagA-seropositivity significantly increased the risk for gastric cancer by 2.28- to 2.87-fold, respectively. Among *H. pylori* infected populations, infection with cagA positive strains further increased the risk by 1.64-fold (95% CI 1.21–2.24) overall and by 2.01-fold for non-cardiac gastric cancer (95% CI 1.21–3.32) [41]. Two years earlier, Ekstrom et al. [35] had published the results of their population-based case-control study involving 298 patients with gastric cancer and 244 controls. The OR of 2.2 (95% CI 1.4–3.6) for non-cardiac gastric cancer in *H. pylori* positive patients rose to 21.0

(95% CI 8.3–53.4) if an additional immunoblot against cagA antibodies that prevail longer after eradication was taken into account [35]. Concerning host genetic factors, specific polymorphisms of the gene for IL-1β, which plays a key role in initiating and amplifying the inflammatory response to the infection with *H. pylori*, significantly increase the risk for pre-cancerous conditions [42]. Additionally, the *H. pylori*-induced inflammation observed in relatives of patients with gastric cancer is more severe than in infected individuals without a positive family history. Whether eradication of the infection reduces the risk of gastric cancer and if there is a 'point of no return' after which eradication no longer has any effect are central issues addressed by clinical studies in the past.

To evaluate the efficacy of *H. pylori* eradication in the prevention of gastric cancer, Fuccio et al. [43] performed a review of the published epidemiological, observational and interventional studies. This meta-analysis of randomized controlled trials taking into account 2,020 patients who underwent eradication therapy and 2,020 controls revealed an OR of 0.67 (95% CI 0.42–1.07) in favour of eradication therapy, showing that *H. pylori* eradication is a primary chemo-preventive strategy of gastric cancer. While Wong et al. [44] were just able to show a beneficial effect of eradication therapy for the prevention of gastric cancer in patients without atrophy, intestinal metaplasia or dysplasia, Fukase et al. [45] showed a beneficial effect of *H. pylori* eradication therapy for the prevention of metachronous gastric cancer even after endoscopic resection of early gastric cancer. They randomized 544 patients with early gastric cancer into 2 treatment arms, with subjects receiving either eradication therapy or standard therapy. In the intention-to-treat analysis the OR for metachronous gastric cancer was 0.353 (95% CI 0.161–0.775, $p = 0.009$) [45].

Currently, clinical data suggest that pre-cancerous abnormalities such as atrophic gastritis and intestinal metaplasia are possibly reversible in some patients after *H. pylori* eradication. Eradication therapy should, therefore, be considered even in patients with pre-cancerous lesions. Nonetheless, the fact that progression of atrophic gastritis and intestinal metaplasia to cancer also takes place after *H. pylori* eradication therapy suggests that there might be a point of no return where irreversible genetic and epigenetic changes have already taken place and the cascade leading to cancer can no longer be interrupted.

A definite answer to the question whether eradication of *H. pylori* can reduce the risk of developing gastric cancer could only be given by a prospective randomized clinical trial. However, such a trial will be impossible to perform as it would require an immense number of participants (more than 100,000) and very long-term follow-up (more than 2 decades).

H. pylori eradication therapy is currently not widely used for the purpose of gastric cancer prevention. The main reasons are the lack of cost-effectiveness in regions of low incidence, the complexity of the mandatory triple therapy combining 2 antibiotics with a PPI and the rise in resistance to antibiotics that would be a consequence.

Another important question to consider is whether the infection with *H. pylori* is a necessary prerequisite for gastric cancer or whether gastric cancer can develop irrespective of the infection. This is the case for cardiac gastric cancer that accounts for up to 20% of gastric cancer in some countries [46], certain hereditary gastric cancers and in those patients in whom the disease results from autoimmune gastritis where infection with *H. pylori* is uncommon [47].

As the diagnosis of gastric cancer is confirmed at an advanced stage of disease in 80% of patients and curative therapy is then no longer an option, primary and secondary prevention in terms of early diagnosis and effective screening measures are all the more important. Endoscopic mass screening for gastric cancer is cost-effective in populations with moderate to high risk [48]. In regions with low incidence, such as Europe, a cheap and effective method to identify individuals with high risk of developing gastric cancer is needed. The measurement of serum pepsinogen I and II in combination with gastrin 17 and IgG antibodies against *H. pylori* is a helpful tool to identify high-risk individuals [49] but is not suited for a routine mass screening. The recently published German S3-guideline '*H. pylori* and gastroduodenal ulcer disease' [29] states that eradication therapy can be conducted in patients at high risk of developing gastric cancer but emphasizes that mass screening is not cost effective in Germany.

MALT Lymphoma

According to the current World Health Organization classification, gastric MALT lymphoma is an extranodal marginal zone B-cell lymphoma of MALT type and belongs to the indolent (formerly low grade) gastric lymphomas that account for 40% of primary gastric lymphomas. In about 90% of cases this lymphoma is associated with *H. pylori* infection. Curing *H. pylori* infection is associated with complete remission of gastric MALT lymphoma in approximately 80% of patients with *H. pylori* positive gastric low-grade MALT-lymphoma in stage I according to the Radaszkiewicz classification [50]. An absence of gene t(11,18) (q21;q21) translocation with fusion of API2 and MLT has been recognized as a predictor of response to eradication therapy. In a long-term follow-up study by Wündisch et al. [51] it was shown that 80% of 120 patients with stage I disease who underwent eradication therapy achieved a complete histologic remission. Eighty percent of them stayed in continuous complete remission after a median follow-up of 75 months. Therefore, *H. pylori* eradication is the treatment of first choice for *H. pylori* infected individuals with stage I low grade MALT lymphoma [4]. As the treatment success is significantly lower in patients with stage II disease, the role of eradication therapy in these patients is still matter of debate.

Extragastric Diseases

The hypothetical role of *H. pylori* in various extragastric diseases has been described in many clinical studies since the first description of the bacterium, but most of these studies are epidemiological with several other variables or observational treatment trials.

The link with idiopathic thrombocytic purpura was first described by Gasbarrini et al. in 1998 [52]. Since then, the link between *H. pylori* infection and idiopathic thrombocytic purpura has been confirmed in studies demonstrating that platelet counts in patients with the condition return to normal after eradication therapy. A recent meta-analysis of clinical studies with 696 patients showed an overall response rate of 50.3% (95% CI 41.6–59%) to *H. pylori* eradication therapy [53]. The favourable effect of eradication therapy was even observed in some cases after long-term follow-up [54].

Iron deficiency anaemia is a further extragastric disease linked to *H. pylori* infection if other possible causes have been excluded. In a recent meta-analysis, Muhsen and Cohen [55] found an OR of 2.8 (95% CI 1.9–4.2) for iron deficiency anaemia in *H. pylori* infected patients. Hershko et al. [56] showed a normalization of haemoglobin levels after eradication therapy in 25 young male patients with unexplained iron deficiency anaemia.

With current knowledge, unexplained Iron deficiency anaemia and idiopathic thrombocytic purpura are recommended indications for *H. pylori* eradication.

The link of *H. pylori* infection with other extragastric diseases, atopic cutaneous and pulmonary diseases are mostly derived from small case series and small pilot studies and need further investigation.

Conclusions

H. pylori is the key pathogen in peptic ulcer disease and gastric cancer. Its eradication must strongly be recommended in infected patients with duodenal or gastric ulcers and MALT lymphoma as well as following endoscopic mucosa resection or endoscopic submucosal dissection for early gastric cancer. *H. pylori* eradication before the development of gastric atrophy or intestinal metaplasia has the best potential to prevent gastric cancer. The challenge for the future will not only be to cure peptic ulcer disease and to prevent gastric cancer by eradication of *H. pylori* but to prevent even the infection which would eliminate the risk for all *H. pylori*-related complications.

References

1 Marshall BJ, Warren JR: Unidentified curved bacilli in the stomach of patients with gastritis and peptic ulceration. Lancet 1984;1:1311–1314.

2 Kuipers EJ: *Helicobacter pylori* and the risk and management of associated diseases: gastritis, ulcer disease, atrophic gastritis and gastric cancer. Aliment Pharmacol Ther 1997;11(suppl):71–88.

3 Correa P: A human model of gastric carcinogenesis. Cancer Res 1988;48:3554–3560.

4 Malfertheiner P, Megraud F, O'Morain C, et al: Current concepts in the management of *Helicobacter pylori* infection: the Maastricht III Consensus Report. Gut 2007;56:772–781.

5 O'Connor HJ, Kanduru C, Bhutta AS, Meehan JM, Feeley KM, Cunnane K: Effect of Helicobacter pylori eradication on peptic ulcer healing. Postgrad Med J 1995;71:90–93.

6 Hunt RH, Bazzoli F: Review article: should NSAID/low-dose aspirin takers be tested routinely for *H. pylori* infection and treated if positive? Implications for primary risk of ulcer and ulcer relapse after initial healing. Aliment Pharmacol Ther Suppl 2004;19:9–16.

7 Huang JQ, Sridhar S, Hunt RH: Role of *Helicobacter pylori* infection and non-steroidal anti-inflammatory drugs in peptic-ulcer disease: a meta-analysis. Lancet 2002;359:14–22.

8 Chan FKL, To KF, Wu JCY, et al: Eradication of *Helicobacter pylori* and risk of peptic ulcers in patients starting long-term treatment with non-steroidal anti-inflammatory drugs: a randomised trial. Lancet 2002;359:9–13.

9 De Leest HTJI, Steen KSS, Lems WF, et al: Eradication of *Helicobacter pylori* does not reduce the incidence of gastroduodenal ulcers in patients on long-term NSAID treatment: double-blind, randomized, placebo-controlled trial. Helicobacter 2007;12:477–485.

10 Labenz J, Blum AL, Bolten WW, et al: Primary prevention of diclofenac associated ulcers and dyspepsia by omeprazole or triple therapy in *Helicobacter pylori* positive patients: a randomised double blind, placebo controlled, clinical trial. Gut 2002;51:329–335.

11 Chan FK, Chung SC, Suen BY, et al: Preventing recurrent upper gastrointestinal bleeding in patients with *Helicobacter pylori* infection who are taking low-dose aspirin or naproxen. N Engl J Med 2001; 344:967–973.

12 De Leest HT, Steen KS, Bloemena E, et al: *Helicobacter pylori* eradication in patients on long-term treatment with NSAIDs reduces the severity of gastritis: a randomized controlled trial. J Clin Gastroenterol 2009;43:140–146.

13 Malfertheiner P, Lind T, Willich S, et al: Prognostic influence of Barrett's oesophagus and *Helicobacter pylori* infection on healing of erosive gastro-oesophageal reflux disease (GORD) and symptom resolution in non-erosive GORD: report from the ProGORD study. Gut 2005;54:746–751.

14 Richter JE: Effect of *Helicobacter pylori* eradication on the treatment of gastrooesophageal reflux disease. Gut 2004;53:310–311.

15 Metz DC, Kroser JA: *Helicobacter pylori* and gastroesophageal reflux disease. Gastroenterol Clin North Am 1999;28:971–985.

16 Pandolfino JE, Howden CW, Kahrilas PJ: *H. pylori* and GERD: is less more? Am J Gastroenterol 2004; 99:1222–1225.

17 Wu JCY, Chan FKL, Wong SKH, Lee YT, Leung WK, Sung JJY: Effect of *Helicobacter pylori* eradication on oesophageal acid exposure in patients with reflux oesophagitis. Aliment Pharmacol Ther 2002; 16:545–552.

18 Labenz J, Blum AL, Bayerdorffer E, Meining A, Stolte M, Borsch G: Curing Helicobacter pylori infection in patients with duodenal ulcer may provoke reflux esophagitis. Gastroenterology 1997;112: 1442–1447.

19 Malfertheiner P, Dent J, Zeijlon L, et al: Impact of *Helicobacter pylori* eradication on heartburn in patients with gastric or duodenal ulcer disease: results from a randomized trial programme. Aliment Pharmacol Ther 2002;16:1431–1442.

20 Moayyedi P, Bardhan C, Young L, Dixon MF, Brown L, Axon ATR: *Helicobacter pylori* eradication does not exacerbate reflux symptoms in gastroesophageal reflux disease. Gastroenterology 2001;121:1120–1126.

21 Harvey RF, Lane JA, Murray LJ, Harvey IM, Donovan JL, Nair P: Randomised controlled trial of effects of *Helicobacter pylori* infection and its eradication on heartburn and gastro-oesophageal reflux: Bristol *Helicobacter* project. Br Med J 2004;328:1417–1419.

22 Kuipers EJ. Proton pump inhibitors and *Helicobacter pylori* gastritis: friends or foes? Basic Clin Pharmacol Toxicol 2006;99:187–194.

23 Genta RM, Lew GM, Graham DY: Changes in the gastric mucosa following eradication of *Helicobacter pylori*. Mod Pathol 1993;6:281–289.

24 Kuipers EJ, Nelis GF, Klinkenberg-Knol EC, et al: Cure of *Helicobacter pylori* infection in patients with reflux oesophagitis treated with long term omeprazole reverses gastritis without exacerbation of reflux disease: results of a randomised controlled trial. Gut 2004;53:12–20.

25 Talley NJ: Chronic peptic ulceration and nonsteroidal anti-inflammatory drugs: more to be said about NSAIDs? Gastroenterology 1992;102:1074–1076.
26 Delaney BC, Moayyedi P, Forman D: Initial management strategies for dyspepsia. Cochrane Database Syst Rev 2003:2.
27 Allison JE, Hurley LB, Hiatt RA, Levin TR, Ackerson LM, Lieu TA: A randomized controlled trial of test-and-treat strategy for *Helicobacter pylori*: clinical outcomes and health care costs in a managed care population receiving long-term acid suppression therapy for physician-diagnosed peptic ulcer disease. Arch Intern Med 2003;163:1165–1171.
28 Gatta L, Ricci C, Tampieri A, Vaira D: Non-invasive techniques for the diagnosis of *Helicobacter pylori* infection. Clin Microbiol Infect 2003;9:489–496.
29 Fischbach W, Malfertheiner P, Hoffmann JC, et al: S3-guideline '*Helicobacter pylori* and gastroduodenal ulcer disease'. Z Gastroenterol 2009;47:68–102.
30 Lassen AT, Pedersen FM, Bytzer P, Schaffalitzky de Muckadell OB: *Helicobacter pylori* test-and-eradicate versus prompt endoscopy for management of dyspeptic patients. Lancet 2000;356:455–460.
31 Klok RM, Arents NLA, De Vries R, et al: Economic evaluation of a randomized trial comparing *Helicobacter pylori* test-and-treat and prompt endoscopy strategies for managing dyspepsia in a primary-care setting. Clin Ther 2005;27:1647–1657.
32 O'Morain C: Role of *Helicobacter* in functional dyspepsia. World J Gastroenterol 2006;12:2677–2680.
33 McCarthy C, Patchett S, Collins RM, Beattie S, Keane C, O'Morain C: Long-term prospective study of *Helicobacter pylori* in nonulcer dyspepsia. Dig Dis Sci 1995;40:114–119.
34 Moayyedi P, Soo S, Deeks J, et al: Eradication of *Helicobacter pylori* for non-ulcer dyspepsia. Cochrane Database Syst Rev 2006:2.
35 Ekstrom AM, Held M, Hansson LE, Engstrand L, Nyrén O: *Helicobacter pylori* in gastric cancer established by CagA immunoblot as a marker of past infection. Gastroenterology 2001;121:784–791.
36 Correa P: Human gastric carcinogenesis: a multistep and multifactorial process. First American Cancer Society Award lecture on cancer epidemiology and prevention. Cancer Res 1992;52:6735–6740.
37 Watanabe T, Tada M, Nagai H, Sasaki S, Nakao M: *Helicobacter pylori* infection induces gastric cancer in mongolian gerbils. Gastroenterology 1998;115:642–648.
38 Honda S, Fujioka T, Tokieda M, Satoh R, Nishizono A, Nasu M: Development of *Helicobacter pylori*-induced gastric carcinoma in Mongolian gerbils. Cancer Res 1998;58:4255–4259.
39 Uemura N, Okamoto S, Yamamoto S, et al: *Helicobacter pylori* infection and the development of gastric cancer. N Engl J Med 2001;345:784–789.
40 Huang JQ, Sridhar S, Chen Y, Hunt RH: Meta-analysis of the relationship between *Helicobacter seropositivity* and gastric cancer. Gastroenterology 1998;114:1169–1179.
41 Huang JQ, Zheng GF, Sumanac K, Irvine EJ, Hunt RH: Meta-analysis of the relationship between caga seropositivity and gastric cancer. Gastroenterology 2003;125:1636–1644.
42 El-Omar EM: The importance of interleukin 1beta in *Helicobacter pylori* associated disease. Gut 2001; 48:743–747.
43 Fuccio L, Zagari RM, Minardi ME, Bazzoli F: Systematic review: *Helicobacter pylori* eradication for the prevention of gastric cancer. Aliment Pharmacol Ther 2007;25:133–141.
44 Wong BCY, Lam SK, Wong WM, et al: *Helicobacter pylori* eradication to prevent gastric cancer in a high-risk region of China: a randomized controlled trial. J Am Med Assoc 2004;291:187–194.
45 Fukase K, Kato M, Kikuchi S, et al: Effect of eradication of *Helicobacter pylori* on incidence of metachronous gastric carcinoma after endoscopic resection of early gastric cancer: an open-label, randomised controlled trial. Lancet 2008;372:392–397.
46 Hansen S, Melby KK, Aase S, Jellum E, Vollset SE: *Helicobacter pylori* infection and risk of cardia cancer and non-cardia gastric cancer: a nested case-control study. Scand J Gastroenterol 1999;34:353–360.
47 Malfertheiner P, Sipponen P, Naumann M, et al: *Helicobacter pylori* eradication has the potential to prevent gastric cancer: a state-of-the-art critique. Am J Gastroenterol 2005;100:2100–2115.
48 Dan YY, So JB, Yeoh KG: Endoscopic screening for gastric cancer. Clin Gastroenterol Hepatol 2006;4: 709–716.
49 Miki K, Urita Y: Using serum pepsinogen wisely in a clinical practice. J Dig Dis 2007;8:8–14.
50 Morgner A, Schmelz R, Thiede C, Stolte M, Miehlke S: Therapy of gastric mucosa associated lymphoid tissue lymphoma. World J Gastroenterol 2007;13: 3554–3566.
51 Wündisch T, Thiede C, Morgner A, et al: Long-term follow-up of gastric MALT lymphoma after Helicobacter pylori eradication. J Clin Oncol 2005;23: 8018–8024.
52 Gasbarrini A, Franceschi F, Tartaglione R, Landolfi R, Pola P, Gasbarrini G: Regression of autoimmune thrombocytopenia after eradication of *Helicobacter pylori*. Lancet 1998;352:878.
53 Stasi R, Sarpatwari A, Segal JB, et al: Effects of eradication of *Helicobacter pylori* infection in patients with immune thrombocytopenic purpura: a systematic review. Blood 2009;113:1231–1240.

54 Satake M, Nishikawa J, Fukagawa Y, et al: The long-term efficacy of *Helicobacter pylori* eradication therapy in patients with idiopathic thrombocytopenic purpura. J Gastroenterol Hepatol 2007;22:2233–2237.
55 Muhsen K, Cohen D: *Helicobacter pylori* infection and iron stores: a systematic review and meta-analysis. Helicobacter 2008;13:323–340.
56 Hershko C, Ianculovich M, Souroujon M: A hematologist's view of unexplained iron deficiency anemia in males: impact of *Helicobacter pylori* eradication. Blood Cells Mol Dis 2007;38:45–53.

Prof. Dr. Dr. Peter Malfertheiner, MD
Department of Gastroenterology, Hepatology and Infectious Diseases
Otto von Guericke University, Magdeburg
Leipziger Strasse 44, DE–39120 Magdeburg (Germany)
Tel. +49 391 671 3100, Fax +49 391 671 3105, E-Mail Peter.Malfertheiner@med.ovgu.de

Mayerle J, Tilg H (eds): Clinical Update on Inflammatory Disorders of the Gastrointestinal Tract.
Front Gastrointest Res. Basel, Karger, 2010, vol 26, pp 199–210

Carcinogenesis and Treatment of Gastric Cancer

Roland Rad[a] · Matthias Ebert[b]

[a]The Wellcome Trust Sanger Institute, Cambridge, UK; Department of Medicine II, Klinikum rechts der Isar, Technische Universität München, Munich, Germany

Abstract

Gastric cancer is the second most common cause of cancer-related death and, despite a decreasing incidence in the USA and Europe, overall it remains one of the most frequent cancers worldwide. *Helicobacter pylori* infection is has been recognized to be the leading cause of both intestinal and diffuse type distal gastric cancer, yet only 1% of colonized individuals develop malignancy. *H. pylori* strain characteristics and host genetics influence the risk of cancer development by differentially affecting the strength and the pattern of the host inflammatory response. Patients usually present with advanced disease, in which a curative treatment approach is often not possible. However, resection of the cancer remains the most important treatment option. In recent years, neoadjuvant treatment approaches have been developed and led to an improved overall survival. The introduction of novel targeted agents is a promising strategy which will further improve the prognosis of patients with advanced tumour stages.

Gastric cancer (GC) is responsible for approximately 10% of all death from cancer worldwide. The age-adjusted incidence of GC shows considerable geographic variation, being as high as 70 per 100,000 men in East Asian countries, such as Japan, but only 10 per 100,000 males in the USA and even much lower in some African countries [1]. Non-cardia (distal) GC is strongly associated with *H. pylori* infection. In contrast, proximal adenocarcinamas of the cardia and the distal oesophagus are inversely associated with *H. pylori*, indicative of a protective effect of infection [2, 3]. The two histological types of distal GC have different epidemiological and pathophysiological features. Intestinal-type gastric adenocarcinoma usually occurs at a late age, predominates in men and progresses through a relatively well-defined series of histological steps. Diffuse-type GC is more common among younger people, affects men and women equally and tends to have a worse prognosis than intestinal-type GC [1, 3].

H. pylori and Gastric Carcinogenesis

H. pylori infection is the most important factor in the aetiology of non-cardia GC [3, 4]. Seropositivity for the bacterium is associated with an up to 16-fold increased risk for distal GC [4, 5] and prospective studies indicate that bacterial eradication decreases the risk of cancer development, at least in the subset of patients that had not developed pre-malignant lesions until treatment [6, 7]. Consistently, the progressive decline of infection in developed countries has been mirrored by a decreasing incidence of distal GC. In many case control studies, the risk of gastric cancer attributable to *H. pylori* has probably been underestimated, because bacterial colonization diminishes in the presence of premalignant lesions, making it difficult to be detected in all cases [5].

Although *H. pylori* is the strongest risk factor for distal GC, only a small percentage of carriers (~1%) develop malignancy. Cancer risk is believed to be influenced by environmental factors, host genetics and *H. pylori* strain differences. *H. pylori* populations are extremely diverse, due to point mutations, substitutions, insertions and/or deletions in their genomes. The pathogenicity of bacteria is determined by the expression of certain virulence factors, such as the type-IV secretion system, the cytotoxin VacA or the adhesin BabA. *H. pylori* uses the type-IV secretion system to inject bacterial macromolecules such as CagA into epithelial cells, thereby inducing oncogenic and pro-inflammatory signal transduction pathways [8]. CagA-positive strains are more common among GC patients than CagA-negative strains, which are rarely disease-associated. In countries with a high GC incidence, such as Japan, virtually all *H. pylori* strains express/secrete BabA/CagA/VacA, whereas in some Western populations the proportion of these 'triple-positive' strains is less than 50% [8, 9]. Furthermore, due to structural modifications, CagA expressed by East-Asian *H. pylori* strains is biologically more active, which further contributes to disease pathogenesis [8].

Inflammation, Atrophy, Metaplasia and Intestinal-Type Gastric Cancer

GC develops as a consequence of chronic gastric inflammation, which progresses to multifocal atrophic gastritis, metaplasia, dysplasia and finally adenocarcinoma (fig. 1) [10]. The host inflammatory response is of key importance in gastric carcinogenesis. This is illustrated by the fact that pre-cancerous gastric changes observed in *Helicobacter*-infected wild-type mice do not occur in T cell-deficient animals [11]. In humans, it is not only the severity of the gastritis, but also its distribution in the stomach that affects the risk of disease. Gastritis that is confined to the antrum is associated with an increased risk of duodenal ulceration. *H. pylori*-infected persons with duodenal ulceration have a reduced risk of developing GC [12, 13]. In contrast, severe corpus-predominant or pan-gastritis predisposes to multifocal atrophy and

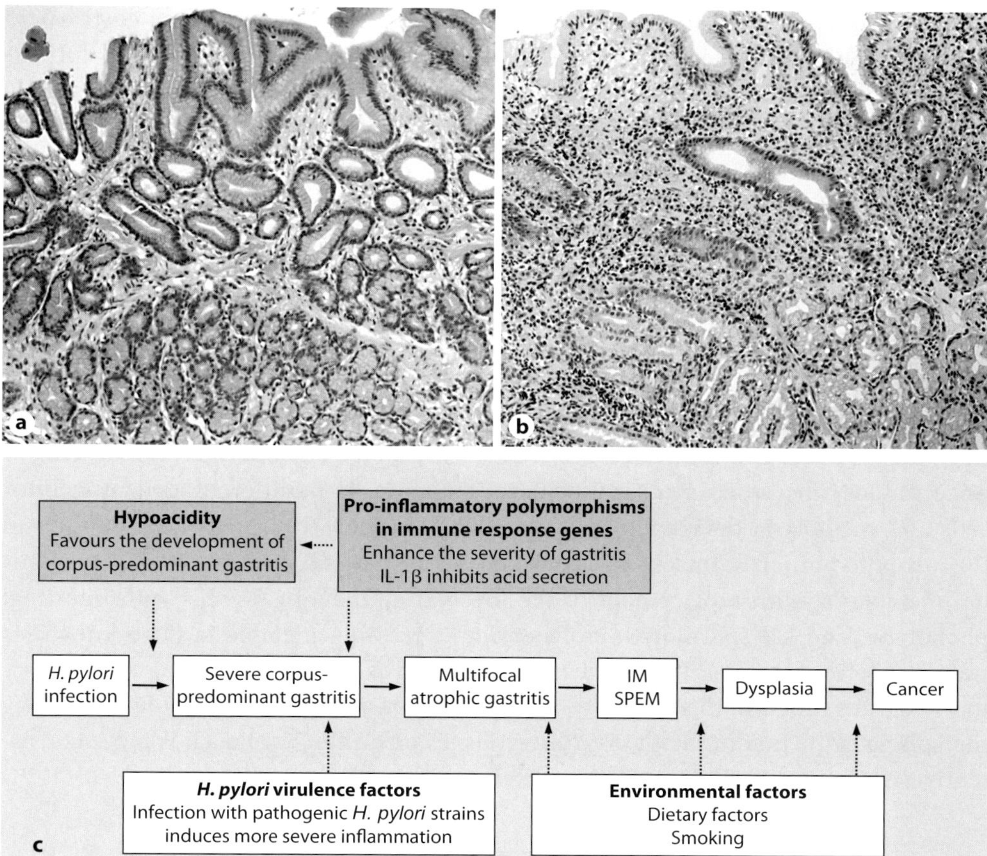

Fig. 1. Progression model of intestinal type distal gastric cancer. Histology of uninfected (**a**) and *H. pylori* infected gastric mucosa with severe inflammation (**b**). **c** Intestinal-type GC develops as a consequence of *H. pylori*-induced chronic gastric inflammation. It is the combination of host genetics (shaded boxes), bacterial virulence factors and environmental insults that drive the natural course of infection toward cancer. High acid secretion predisposes to the development of antrum-predominant gastritis leading to an increased risk of duodenal ulceration, but rarely cancer (not depicted). In contrast, hypoacidity predisposes to corpus-predominant gastritis, which is linked to cancer development. The severity of the inflammatory response depends on infection with more pathogenic *H. pylori* strains, which possess certain virulence factors such as the bacterial type-IV secretion system, the adhesin BabA or the cytotoxin VacA. The degree of inflammation can further be influenced by pro-inflammatory polymorphisms in human immune response genes (e.g. *IL1B, IL10, TNFA, IL8, TLR4*). Increased IL-1β expression in persons harbouring pro-inflammatory *IL-1* polymorphisms can also lead to inhibition of gastric acid secretion. The development of multifocal atrophic gastritis with parietal cell loss leads to the development of intestinal metaplasia (IM) and/or spasmolytic polypeptide expressing metaplasia (SPEM; pseudopyloric metaplasia), which can progress to dysplasia and cancer. Environmental factors, such as smoking and dietary factors (e.g. high-salt diet, low fruit or vegetable intake, high consumption of processed meat) can influence cancer development at multiple stages.

GC development. The pattern of gastritis is influenced by the extent of gastric acid secretion, with high or low acid secretion being associated with antrum- and corpus-predominant gastritis, respectively. The reasons for these observations are not entirely understood. One hypothesis is that in hosts with hypoacidity, *Helicobacter* infection might spread more easily from the antrum (where the pH is lower and infection is usually first established) to the corpus [13].

Multifocal atrophy is the most reliable histologic correlate with GC [10]. Patients with severe multifocal atrophic gastritis have an up to 90-fold increased risk of developing adenocarcinoma. The loss of parietal cells in the atrophic stomach leads to severe disruption of glandular homeostasis. Parietal cells are the major organizing centres of fundic units, orchestrating complex programs of cellular growth and differentiation through the secretion of important paracrine regulators such as sonic hedgehog and TGF-α. Drug-induced or genetically engineered ablation of parietal cells in mice induces increased proliferation, accumulation of undifferentiated progenitor cells and metaplasia development. Two types of mucous cell metaplasia develop in the atrophic human stomach as a consequence of parietal cell loss [10]: (1) intestinal metaplasia, with replacement of the foveolar epithelium by cells with intestinal phenotype, and (2) spasmolytic polypeptide-expressing metaplasia (also known as pseudopyloric metaplasia), defined as the presence of glands with antral morphology in the corpus mucosa. In several animal models, spasmolytic polypeptide-expressing metaplasia is the predominant pre-cancerous lesion and in humans it is more consistently associated with GC than intestinal metaplasia [10].

Mechanisms Linking Inflammation and Cancer

Despite evidence for the key role of inflammation in gastric carcinogenesis, the mechanisms that link infection, inflammation and cancer are only partly understood. Cancer is believed to arise from malignant transformation of pluripotent gastric or bone marrow-derived stem cells through the accumulation of multiple genetic defects. Chronic inflammation and tissue damage are triggers for increased proliferation and the accumulation of undifferentiated progenitor cells. These cells are exposed to mutagenic substances, such as reactive oxygen and nitrogen species, which are released from inflammatory cells. Important components in the linkage between inflammation and cancer are inflammatory cell-derived cytokines and growth factors, which dysregulate multiple processes of cellular and mucosal homeostasis, such as proliferation, apoptosis, angiogenesis, invasiveness and metastasis. *H. pylori* induces a T helper type-1 cellular immune response in the stomach, characterized by the presence of specific cytokines, such as IFN-γ, IL-12, TNF-α and IL-1β [4]. Cancer development is linked to this Th1-specific gastric microenvironment. In *Helicobacter*-infected mice experimental induction of a Th2 response attenuates the development of atrophy [14]. Gastric cancer rates in different parts of the world correlate with the strength of Th1 polarization of the antibacterial immune response in different populations, supporting the Th1/Th2 paradigm [10, 15]. Accordingly, GC risk is influenced by variations in Th1 cytokine genes. Pro-inflammatory polymorphisms in the human *IL-1B* and IL-1 receptor antagonist *(IL-1RA)* genes, for example, have been associated with increased gastric IL-1β levels and a highly increased cancer risk, particularly in persons infected with virulent *H. pylori* strains [16, 17]. The importance of IL-1β in gastric carcinogenesis is further underscored by the observation that transgenic mice over-expressing IL-1β develop gastric malignancy [18]. The cancer-promoting effects of IL-1β are mediated by both its pro-inflammatory and antisecretory effects. IL-1β is a very potent inhibitor of gastric acid secretion and therefore facilitates the development of corpus-predominant gastritis [16].

Molecular Alterations and Oncogenic Pathways in Gastric Cancer

Only 1–3% of all GC cases are related to hereditary GC predisposition syndromes [19]. Germline mutations of *CDH1* (E-cadherin) are responsible for about one third of hereditary diffuse GCs, whereas hereditary non-polyposis colorectal cancer syndrome (*hMLH1/hMSH2* mutations) increases the risk of developing mostly intestinal-type GCs. Furthermore, persons affected by familial adenomatous polyposis (*APC* mutations), Peutz-Jeghers syndrome (*LKB1* mutations), Li-Fraumeni syndrome (*TP53* mutations) and germ-line mutations in *BRCA2* have an increased risk of GC.

A large number of genes and pathways regulating cell growth and proliferation have been found to be altered in sporadic GC [19], but only a few with frequent alterations have been reported and will be mentioned here [reviewed in more detail in 19]. Amplifications of the growth factor receptor genes *ERBB2* (HER2/NEU) or *MET* are each present in about 10% of both histological GC types. Other tyrosine kinase growth factor receptors, such as EGFR, PDGFRA, VEGFR and FGFR have been reported to be frequently up-regulated in GC, mainly by over-expression. Aberrant growth factor signalling can also occur due to mutations in downstream signalling components of the tyrosine kinase receptor pathway, such as KRAS and PI3K, which have been found to have oncogenic mutations in up to 10% of GC cases [19]. Alterations of another important growth regulatory signalling pathway, the TGF-β pathway, can result from mutations or epigenetic silencing of the type-I or type-II TGF-β receptors or the downstream signalling molecules SMAD3/SMAD4 [19]. TGF-β receptor inactivation can occur as a consequence or microsatellite instability, which is observed in 5–40% of GCs, mostly due to *hMLH1* hypermethylation. Furthermore, deregulations in cytokine signalling (e.g. increased STAT3 activation) and alterations in the Wnt pathway are considered to be involved in gastric carcinogenesis, although APC or β-Catenin mutations are far less common than in colorectal cancer. Somatic mutations in E-cadherin are found specifically in sporadic diffuse-type GCs (33–50%) whereas *TP53* mutations frequently occur in both histological types of GC (in up to 40% of GC cases). In addition, *P16* expression is frequently down-regulated (25–42%) by promoter methylation, whereas c-myc levels are increased (15%) due to over-expression or gene amplification [19].

Treatment of Gastric Cancer

Overall prognosis for patients with GC is poor. This results mainly from the advanced tumour stage at the time of diagnosis. Most patients present either with locally advanced disease or with distant metastasis. In these cases a curative treatment strategy is mostly not possible. This results in an overall 5-year survival rate of 10–25%, depending on the region in which GC is treated. In patients with early GC, resection remains the most important option. In special cases, i.e. cancers limited to the mucosa, endoscopic resection with close follow-up may be possible. However, in cancers of the diffuse type, with ulcers and with a size of more than 2 cm surgical resection should be performed. In as much as the nodal involvement is the single most important prognostic factor in this disease, the role of lymph node dissection has been extensively analysed. While the D2 dissection is now considered the standard of surgical resection in GC, recent studies with an even more aggressive approach, termed D3 dissection, did not lead to an improved outcome in patients with advanced GC [20]. In order to improve outcome among these patients, multimodal treatment strategies have been developed. These have led to an improvement in overall survival, and they should now be regarded as standard of care in patients with locally advanced GC.

Neoadjuvant Therapy

The aim of neoadjuvant therapy is to down-stage the primary tumour in order to improve the R0 resection rate. This strategy has been successfully implemented in the treatment of various cancers, including rectal and GC. While in the past the value of neoadjuvant therapy in GC was still under debate, 2 recent phase III studies have now confirmed its role in the treatment of GC. The famous MAGIC trial was recently published by Cunningham et al. [21]. In this study patients with stage II and III GC and oesophageal adenocarcinoma were randomized to either perioperative chemotherapy with epirubicin, cisplatin and 5-FU and surgery versus surgery alone. The final results showed that this pre- and post-operative treatment resulted in a 13% improvement of overall survival at 5 years. Interestingly, this improvement of outcome was observed in all subgroups of patients, i.e. patients with oesophageal cancers, cardia cancers and GCs. These results were recently confirmed by the FNCC/FFCD phase III study in which patients with stage II and III gastric and gastro-oesophageal junction cancers were randomized to perioperative treatment with cisplatin and 5-FU and surgery versus surgery alone. Again, the additional chemotherapy resulted in a 12% improvement in 5-year survival (HR 0.69, 95% CI 0.50–0.95, p = 0.021). While these 2 studies have independently confirmed the value of additional neoadjuvant (and possibly adjuvant chemotherapy in GC, a further improvement in prognosis in this group of patients is still warranted. Two different strategies have been developed in order to further improve this outcome. In the recently initiated MAGIC-B (ST03) study, the study group set up a similar treatment design as in the MAGIC study; however, the VEGF-targeting antibody bevacizumab was added to the chemotherapy regimen. This strategy follows the current trend to improve the efficacy of treatment by combining chemotherapy with targeted agents. A different approach has been developed by the Munich group. Preliminary studies performed in patients with oesophgeal and proximal GCs have shown that treatment response can be predicted by functional PET imaging [22]. Using a FDG-tracer the uptake in the cancer cell can be assessed by PET imaging. Furthermore, the decrease in FDG uptake early, after 2 weeks of chemotherapy, allows the assessment of treatment response and efficacy and thus is helpful in the selection of patients who will benefit from neoadjuvant chemotherapy versus individuals without response who should no longer be considered as candidates for further chemotherapy. With the help of early treatment response assessment by FDG-PET imaging Lordick et al. [23] designed a trial in which patients underwent pre-operative chemotherapy. PET imaging was performed prior to chemotherapy and again 2 weeks after. Patients with a >35% decrease in glucose uptake after 2 weeks were considered as responders and treatment was continued for 12 weeks, followed by resection of the cancer. In patients with a decrease in uptake <35%, chemotherapy was stopped and resection was performed immediately. This treatment strategy resulted in a significant improvement of survival for metabolic responders, while patients without metabolic response showed a poor survival [23]. This selection strategy of individualized therapy of cancers based on metabolic response is a novel approach which is under further assessment. In the subsequent MUNICON II trial treatment was modified in that metabolic non-responders received additional radiation therapy in order to improve outcome. The results of this study are not yet available.

Adjuvant Therapy

The single most important prognostic factor regarding survival in patients with locally advanced GC is nodal involvement. Multiple studies have confirmed that the presence of lymph node metastasis and a greater degree of nodal involvement is associated with poor survival (fig. 2). Apart from the R status of resection, the presence of distant metastasis and the primary T stage of the tumour, the ratio

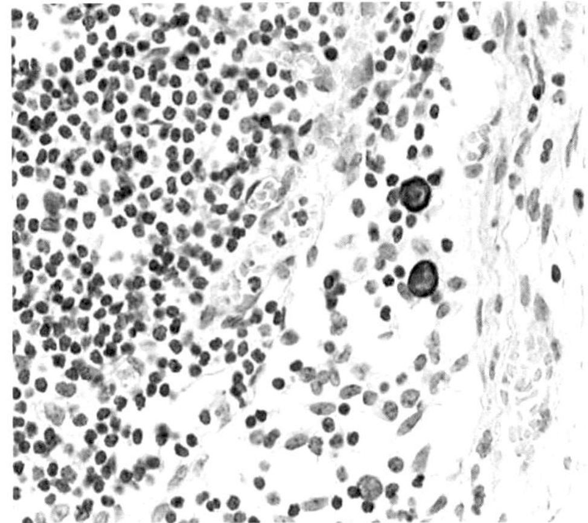

Fig. 2. Detection of infiltrating cancer cells in lymph node metastasis.

between positive and total number of removed lymph node metastasis determines the subsequent course of the disease [24]. While there is overall agreement regarding the role of nodal involvement and its impact on survival, the role of adjuvant therapy in patients with R0 resection and positive lymph nodes is still under debate.

The landmark study INT-0116 assessed the role of adjuvant chemoradiation in GC. Patients undergoing R0 resection for GC or gastro-oesophageal junction cancers were randomized to either post-operative chemoradiation or no adjuvant therapy [25]. In this study patients with adjuvant chemoradiation showed a significant improvement in overall survival compared to patients undergoing surgery only. Today, this treatment regimen is considered standard of care in USA after resection. However, the trial has been widely criticized for the poor performance of the surgical arm. Many patients did not receive the required nodal dissection and, thus, the surgical results are considered not standard and below the expected performance. The role of adjuvant chemotherapy is also still largely unknown. Several meta-analyses have assessed the performance of different adjuvant treatment protocols in patients after GC resection (table 1). However, most of them have found only a modest improvement of survival (if any at all) and, therefore, adjuvant chemotherapy cannot be generally recommended outside clinical trials. Debate has restarted after the publication of the study by Sakuaramoto et al. [26] in which patients after gastrectomy and D2 dissection were randomized either to adjuvant chemotherapy with S1 or follow-up only. S1 is an oral 5-FU prodrug which includes a DPD inhibitor and an inhibitor of 5-FU phosporylation at a ratio of 1:04:1. Treatment was given at a dose of 40 mg/m^2 b.i.d. over 12 months. In this study, treatment with S1 improved significantly overall survival at 3 years (HR 0.8, 95% CI 0.52–0.87, $p = 0.003$), as well as recurrence-free survival (HR 0.50, 95% CI 0.50–0.77, $p < 0.001$) [26]. While these results cannot be

Table 1. Meta-analysis of adjuvant chemotherapy in GC

Meta-analysis	Studies, n	Patients, n	OR (95% CI)	Result
Hermans et al. [31]	11	2,096	0.88 (0.78–1.08)	no benefit
Earle and Maroun [32]	13	1,990	0.80 (0.66–0.97)	small benefit
Mari et al. [33]	21	3,658	0.82 (0.75–0.89)	small benefit
Janunger et al. [34]	21	3,962	0.84 (0.74–0.96)	possible benefit

generalized for all regions of the world, the findings should be carefully analysed and should be used for the design of a similar study in other regions in order to assess the role of adjuvant chemotherapy in patients undergoing gastrectomy and standardized D2 dissection outside Asia.

Palliative Chemotherapy

In the past, multiple chemotherapies have been used in the treatment of advanced GC. The regimens often used FAMTX or ELF, among other combinations. However, recently it has been generally acknowledged that for first-line treatment of GC a combination using a platin compound and/or a fluoropyrimidine and/or taxanes and/or a topoisomerase inhibitor can be used. Mostly either a combination of cisplatin (or oxaliplatin) with 5-FU is used, and can be combined with docetaxel. In a recent study by Lutz et al. [27] the role of PLF as the main treatment for advanced GC has been confirmed. Also the combination of docetaxel, cisplatin and 5-FU is an approved treatment for patients with advanced GC [28]. In a recent study oral 5-FU prodrugs, such as capecitabine, were shown to be of similar efficacy in the treatment of GC [29]. Also, the REAL-2 study in which different platin compounds and fluoropyrimidines were compared revealed that these treatment approaches are largely similar with regard to their performance [30]. The V325 study group analysed the role of additional docetaxel in this combination strategy. In their study DCF was superior with regard to time to progression and overall survival versus the standard cisplatin and 5-FU treatment [28]. However, this increased efficacy was associated with a higher rate of neutropenia and neutropenic infection, thus indicating that further combinations may lead to increased toxicity. Therefore, current treatment strategies are aimed at the assessment of novel targeted therapies, including EGFR inhibitors and antibodies, as well as VEGF antibodies. While several phase II studies show interesting and promising preliminary results, the phase III studies are still ongoing and definite findings are not yet available. In the case of progression under first-line therapy, novel

Table 2. Summary of recent second-line trials in advanced in GC

Author	Phase	Patients, n	Therapy	RR	Time to progression	Overall survival, months
Rosati et al. [35]	II	28	docetaxel 60 mg, capecitabine 2000 mg/m^2	29%	4	6
Kodera et al. [36]	II	44	paclitaxel 80 mg	47%	2.6	7.8
Barone et al. [37]	II	38	docetaxel 75 mg, oxaliplatin 80 mg	47%	4	8.1
Ueda et al. [38]	II	28	irinotecan 70 mg, cisplatin 80 mg	28%	3.5	9.4
Kim et al. [39]	II	57	irinotecan 150 mg, 5FU/leucovorin	21%	2.5	7.6

treatment strategies in a second-line phase are currently under investigation (table 2). Multiple phase II studies assessing the role of novel targeted agents, such as sunitinib, sorafenib, HDAC inhibitors and others, are currently recruiting patients in order to assess the efficacy of these agents in the second-line treatment of advanced GC.

References

1 Forman D, Burley VJ: Gastric cancer: global pattern of the disease and an overview of environmental risk factors. Best Pract Res Clin Gastroenterol 2006;20:633–649.

2 Kamangar F, Dawsey SM, Blaser MJ, Perez-Perez GI, Pietinen P, Newschaffer CJ, Abnet CC, Albanes D, Virtamo J, Taylor PR: Opposing risks of gastric cardia and noncardia gastric adenocarcinomas associated with *Helicobacter pylori* seropositivity. J Natl Cancer Inst 2006;98:1445–1452.

3 Peek RM Jr, Blaser MJ: Helicobacter pylori and gastrointestinal tract adenocarcinomas. Nat Rev Cancer 2002;2:28–37.

4 Suerbaum S, Michetti P: *Helicobacter pylori* infection. N Engl J Med 2002;347:1175–1186.

5 Ekstrom AM, Held M, Hansson LE, Engstrand L, Nyren O: *Helicobacter pylori* in gastric cancer established by CagA immunoblot as a marker of past infection. Gastroenterology 2001;121:784–791.

6 Uemura N, Okamoto S, Yamamoto S, Matsumura N, Yamaguchi S, Yamakido M, Taniyama K, Sasaki N, Schlemper RJ: *Helicobacter pylori* infection and the development of gastric cancer. N Engl J Med 2001;345:784–789.

7 Wong BC, Lam SK, Wong WM, Chen JS, Zheng TT, Feng RE, Lai KC, Hu WH, Yuen ST, Leung SY, Fong DY, Ho J, Ching CK, Chen JS: *Helicobacter pylori* eradication to prevent gastric cancer in a high-risk region of China: a randomized controlled trial. JAMA 2004;291:187–194.

8 Hatakeyama M: Oncogenic mechanisms of the *Helicobacter pylori* CagA protein. Nat Rev Cancer 2004;4:688–694.

9 Rad R, Gerhard M, Lang R, Schoniger M, Rosch T, Schepp W, Becker I, Wagner H, Prinz C: The *Helicobacter pylori* blood group antigen-binding adhesin facilitates bacterial colonization and augments a nonspecific immune response. J Immunol 2002;168: 3033–3041.

10 Fox JG, Wang TC: Inflammation, atrophy, and gastric cancer. J Clin Invest 2007;117:60–69.

11 Roth KA, Kapadia SB, Martin SM, Lorenz RG: Cellular immune responses are essential for the development of *Helicobacter felis*-associated gastric pathology. J Immunol 1999;163:1490–1497.

12 Hansson LE, Nyren O, Hsing AW, Bergstrom R, Josefsson S, Chow WH, Fraumeni JF Jr, Adami HO: The risk of stomach cancer in patients with gastric or duodenal ulcer disease. N Engl J Med 1996;335: 242–249.

13 Amieva MR, El-Omar EM: Host-bacterial interactions in *Helicobacter pylori* infection. Gastroenterology 2008;134:306–323.

14 Smythies LE, Waites KB, Lindsey JR, Harris PR, Ghiara P, Smith PD: *Helicobacter pylori*-induced mucosal inflammation is Th1 mediated and exacerbated in IL-4, but not IFN-gamma, gene-deficient mice. J Immunol 2000;165:1022–1029.

15 Mitchell HM, Ally R, Wadee A, Wiseman M, Segal I: Major differences in the IgG subclass response to *Helicobacter pylori* in the first and third worlds. Scand J Gastroenterol 2002;37:517–522.

16 El-Omar EM, Carrington M, Chow WH, McColl KE, Bream JH, Young HA, Herrera J, Lissowska J, Yuan CC, Rothman N, Lanyon G, Martin M, Fraumeni JF Jr, Rabkin CS: Interleukin-1 polymorphisms associated with increased risk of gastric cancer. Nature 2000;404:398–402.

17 Rad R, Prinz C, Neu B, Neuhofer M, Zeitner M, Voland P, Becker I, Schepp W, Gerhard M: Synergistic effect of *Helicobacter pylori* virulence factors and interleukin-1 polymorphisms for the development of severe histological changes in the gastric mucosa. J Infect Dis 2003;188:272–281.

18 Tu S, Bhagat G, Cui G, Takaishi S, Kurt-Jones EA, Rickman B, Betz KS, Penz-Oesterreicher M, Bjorkdahl O, Fox JG, Wang TC: Overexpression of interleukin-1beta induces gastric inflammation and cancer and mobilizes myeloid-derived suppressor cells in mice. Cancer Cell 2008;14:408–419.

19 Stock M, Otto F: Gene deregulation in gastric cancer. Gene 2005;360:1–19.

20 Sasako M, Sano T, Yamamoto S, Kurokawa Y, Nashimoto A, Kurita A, Hiratsuka M, Tsujinaka T, Kinoshita T, Arai K, Yamamura Y, Okajima K: D2 lymphadenectomy alone or with para-aortic nodal dissection for gastric cancer. N Engl J Med 2008; 359:453–462.

21 Cunningham D, Allum WH, Stenning SP, Thompson JN, Van d, V, Nicolson M, Scarffe JH, Lofts FJ, Falk SJ, Iveson TJ, Smith DB, Langley RE, Verma M, Weeden S, Chua YJ, MAGIC TP: Perioperative chemotherapy versus surgery alone for resectable gastroesophageal cancer. N Engl J Med 2006;355: 11–20.

22 Herrmann K, Walch A, Balluff B, Tanzer M, Hofler H, Krause BJ, Schwaiger M, Friess H, Schmid RM, Ebert MP: Proteomic and metabolic prediction of response to therapy in gastrointestinal cancers. Nat Clin Pract Gastroenterol Hepatol 2009;6:170–183.

23 Lordick F, Ott K, Krause BJ, Weber WA, Becker K, Stein HJ, Lorenzen S, Schuster T, Wieder H, Herrmann K, Bredenkamp R, Hofler H, Fink U, Peschel C, Schwaiger M, Siewert JR: PET to assess early metabolic response and to guide treatment of adenocarcinoma of the oesophagogastric junction: the MUNICON phase II trial. Lancet Oncol 2007; 8:797–805.

24 Siewert JR, Bottcher K, Stein HJ, Roder JD: Relevant prognostic factors in gastric cancer: ten-year results of the German Gastric Cancer Study. Ann Surg 1998;228:449–461.

25 Macdonald JS, Smalley SR, Benedetti J, Hundahl SA, Estes NC, Stemmermann GN, Haller DG, Ajani JA, Gunderson LL, Jessup JM, Martenson JA: Chemoradiotherapy after surgery compared with surgery alone for adenocarcinoma of the stomach or gastroesophageal junction. N Engl J Med 2001; 345:725–730.

26 Sakuramoto S, Sasako M, Yamaguchi T, Kinoshita T, Fujii M, Nashimoto A, Furukawa H, Nakajima T, Ohashi Y, Imamura H, Higashino M, Yamamura Y, Kurita A, Arai K: Adjuvant chemotherapy for gastric cancer with S-1, an oral fluoropyrimidine. N Engl J Med 2007;357:1810–1820.

27 Lutz MP, Wilke H, Wagener DJ, Vanhoefer U, Jeziorski K, Hegewisch-Becker S, Balleisen L, Joossens E, Jansen RL, Debois M, Bethe U, Praet M, Wils J, Van CE: Weekly infusional high-dose fluorouracil (HD-FU), HD-FU plus folinic acid (HD-FU/FA), or HD-FU/FA plus biweekly cisplatin in advanced gastric cancer: randomized phase II trial 40953 of the European Organisation for Research and Treatment of Cancer Gastrointestinal Group and the Arbeitsgemeinschaft Internistische Onkologie. J Clin Oncol 2007;25:2580–2585.

28 Ajani JA, Moiseyenko VM, Tjulandin S, Majlis A, Constenla M, Boni C, Rodrigues A, Fodor M, Chao Y, Voznyi E, Marabotti C, Van CE: Clinical benefit with docetaxel plus fluorouracil and cisplatin compared with cisplatin and fluorouracil in a phase III trial of advanced gastric or gastroesophageal cancer adenocarcinoma: the V-325 Study Group. J Clin Oncol 2007;25:3205–3209.

29 Kang YK, Kang WK, Shin DB, Chen J, Xiong J, Wang J, Lichinitser M, Guan Z, Khasanov R, Zheng L, Philco-Salas M, Suarez T, Santamaria J, Forster G, McCloud PI: Capecitabine/cisplatin versus 5-fluorouracil/cisplatin as first-line therapy in patients with advanced gastric cancer: a randomised phase III noninferiority trial. Ann Oncol 2009;20:666–673.

30 Cunningham D, Starling N, Rao S, Iveson T, Nicolson M, Coxon F, Middleton G, Daniel F, Oates J, Norman AR: Capecitabine and oxaliplatin for advanced esophagogastric cancer. N Engl J Med 2008;358:36–46.

31 Hermans J, Bonenkamp JJ, Boon MC, Bunt AM, Ohyama S, Sasako M, Van de Velde CJ: Adjuvant therapy after curative resection for gastric cancer: meta-analysis of randomized trials. J Clin Oncol 1993;11:1441–1447.

32 Earle CC, Maroun JA: Adjuvant chemotherapy after curative resection for gastric cancer in non-Asian patients: revisiting a meta-analysis of randomised trials. Eur J Cancer 1999;35:1059–1064.

33 Mari E, Floriani I, Tinazzi A, Buda A, Belfiglio M, Valentini M, Cascinu S, Barni S, Labianca R, Torri V: Efficacy of adjuvant chemotherapy after curative resection for gastric cancer: a meta-analysis of published randomised trials: a study of the GISCAD (Gruppo Italiano per lo Studio dei Carcinomi dell'Apparato Digerente). Ann Oncol 2000;1:837–843.

34 Janunger KG, Hafstrom L, Glimelius B: Chemotherapy in gastric cancer: a review and updated meta-analysis. Eur J Surg 2002;168:597–608.

35 Rosati G, Bilancia D, Germano D, Dinota A, Romano R, Reggiardo G, Manzione L: Reduced dose intensity of docetaxel plus capecitabine as second-line palliative chemotherapy in patients with metastatic gastric cancer: a phase II study. Ann Oncol 2007;(18 suppl 6):vi128–vi132.

36 Kodera Y, Ito S, Mochizuki Y, Fujitake S, Koshikawa K, Kanyama Y, Matsui T, Kojima H, Takase T, Ohashi N, Fujiwara M, Sakamoto J, Akimasa N: A phase II study of weekly paclitaxel as second-line chemotherapy for advanced gastric cancer (CCOG0302 study). Anticancer Res 2007;27:2667–2671.

37 Barone C, Basso M, Schinzari G, Pozzo C, Trigila N, D'Argento E, Quirino M, Astone A, Cassano A: Docetaxel and oxaliplatin combination in second-line treatment of patients with advanced gastric cancer. Gastric Cancer 2007;10:104–111.

38 Ueda S, Hironaka S, Boku N, Fukutomi A, Yoshino T, Onozawa Y: Combination chemotherapy with irinotecan and cisplatin in pretreated patients with unresectable or recurrent gastric cancer. Gastric Cancer 2006;9:203–207.

39 Kim ST, Kang WK, Kang JH, Park KW, Lee J, Lee SH, Park JO, Kim K, Kim WS, Jung CW, Park YS, Im YH, Park K: Salvage chemotherapy with irinotecan, 5-fluorouracil and leucovorin for taxane- and cisplatin-refractory, metastatic gastric cancer. Br J Cancer 2005;92:1850–1854.

Dr. Roland Rad
The Wellcome Trust Sanger Institute, Genome Campus
Cambridge CB10 1SA (UK)
Tel. +44 1223 834 244, Fax +44 1223 494 714, E-Mail RR6@sanger.ac.uk

Author Index

Subject Index